MODELING NATURE

Richard J. Gaylord   Kazume Nishidate

MODELING NATURE

Cellular Automata Simulations
with *Mathematica*®

Richard J. Gaylord
Dept. of Materials Science
University of Illinois
Urbana–Champaign
Urbana, IL 61801 USA

Kazume Nishidate
Dept. of Electrical and Electronic Engineering
Faculty of Engineering
Iwate University
Morioka 020 JAPAN

Publisher: Allan M. Wylde
Publishing Associate: Keisha Sherbecoe
Product Manager: Walter Borden
Production Editor: Robert Wexler
Manufacturing Supervisor: Joe Quatela

Library of Congress Cataloging-in-Publication Data
Gaylord, Richard J.
    Modeling nature: cellular automata simulations with Mathematica /
Richard J. Gaylord, Kazume Nishidate.
        p.   cm.
    Includes bibliographical references (p.   –   ) and index.
    ISBN 0-387-94620-9 (soft : alk. paper)
    1. Cellular automata.   2. Mathematical models.   3. Mathematica
(Computer program language)   I. Nishidate, Kazume.   II. Title.
QA267.5.C45G39   1996
501'.183—dc20   96-5727

Printed on acid-free paper.

© 1996 Springer-Verlag New York, Inc.
Published by TELOS®, The Electronic Library of Science, Santa Clara, California.
TELOS® is an imprint of Springer-Verlag New York, Inc.

Photocomposed pages prepared in TEX by Integre Technical Publishing Company, Inc., Albuquerque, NM.
Printed and bound by Hamilton Printing Co., Rensselaer, NY.
Printed in the United States of America.

9 8 7 6 5 4 3 2 1

ISBN 0-387-94620-9   Springer-Verlag   New York   Berlin   Heidelberg   SPIN 10520311

THE
ELECTRONIC
LIBRARY
OF
SCIENCE

TELOS, The Electronic Library of Science, is an imprint of Springer-Verlag New York with publishing facilities in Santa Clara, California. Its publishing program encompasses the natural and physical sciences, computer science, mathematics, economics, and engineering. All TELOS publications have a computational orientation to them, as TELOS' primary publishing strategy is to wed the traditional print medium with the emerging new electronic media in order to provide the reader with a truly interactive multimedia information environment. To achieve this, every TELOS publication delivered on paper has an associated electronic component. This can take the form of book/diskette combinations, book/CD-ROM packages, books delivered via networks, electronic journals, newsletters, plus a multitude of other exciting possibilities. Since TELOS is not committed to any one technology, any delivery medium can be considered. We also do not foresee the imminent demise of the paper book, or journal, as we know them. Instead we believe paper and electronic media can coexist side-by-side, since both offer valuable means by which to convey information to consumers.

The range of TELOS publications extends from research level reference works to textbook materials for the higher education audience, practical handbooks for working professionals, and broadly accessible science, computer science, and high technology general interest publications. Many TELOS publications are interdisciplinary in nature, and most are targeted for the individual buyer, which dictates that TELOS publications be affordably priced.

Of the numerous definitions of the Greek word "telos," the one most representative of our publishing philosophy is "to turn," or "turning point." We perceive the establishment of the TELOS publishing program to be a significant step forward towards attaining a new plateau of high quality information packaging and dissemination in the interactive learning environment of the future. TELOS welcomes you to join us in the exploration and development of this exciting frontier as a reader and user, an author, editor, consultant, strategic partner, or in whatever other capacity one might imagine.

TELOS, The Electronic Library of Science
Springer-Verlag Publishers
3600 Pruneridge Avenue, Suite 200
Santa Clara, CA 95051

THE ELECTRONIC LIBRARY OF SCIENCE

TELOS Diskettes

Unless otherwise designated, computer diskettes packaged with TELOS publications are 3.5″ high-density DOS-formatted diskettes. They may be read by any IBM-compatible computer running DOS or Windows. They may also be read by computers running NEXTSTEP, by most UNIX machines, and by Macintosh computers using a file exchange utility.

In those cases where the diskettes require the availability of specific software programs in order to run them, or to take full advantage of their capabilities, then the specific requirements regarding these software packages will be indicated.

TELOS CD-ROM Discs

For buyers of TELOS publications containing CD-ROM discs, or in those cases where the product is a stand-alone CD-ROM, it is always indicated on which specific platform, or platforms, the disc is designed to run. For example, Macintosh only; Windows only; cross-platform, and so forth.

TELOSpub.com (Online)

Interact with TELOS online via the Internet by setting your World-Wide-Web browser to the URL: *http://www.telospub.com*.

The TELOS Web site features new product information and updates, an online catalog and ordering, samples from our publications, information about TELOS, data-files related to and enhancements of our products, and a broad selection of other unique features. Presented in hypertext format with rich graphics, it's your best way to discover what's new at TELOS.

TELOS also maintains these additional Internet resources:

gopher://gopher.telospub.com
ftp://ftp.telospub.com

For up-to-date information regarding TELOS online services, send the one-line e-mail message:

send info to: info@TELOSpub.com.

To Calvin and Hobbes: For their profound insights into just about everything. And for making me laugh.

-RJG

To Hiroe. Without her support and patience, this project would never have been possible.

-KN

The What, Why, and How of This Book

What is the Philosophy Behind Cellular Automata Modeling?

A cellular automaton (CA) model is essentially a discrete space-time-state representation of a system of many objects that simultaneously interact with other nearby objects (see chapter 1 for a more detailed specification of a CA).

The sine qua non of a CA model is that the behavior of the model is "determined not by some centralized authority but by local interactions among decentralized components" (see the excellent book *Turtles, Termites and Traffic Jams* (Resnick, 1994) for a thought-provoking discussion of decentralization and the centralized mindset).

Why Use Mathematica for Cellular Automata Modeling?

There are a number of features of the *Mathematica* programming language (see the Appendix and Chapter 1 for more details) that make it especially appropriate for the task of writing CA programs:

- *Mathematica* is fundamentally a term rewriting system (TRS) based on pattern matching.

The use of pattern-matched rewrite rules allows us to express the functions, known as CA rules, that make up the main part of a CA program, directly and easily.

- *Mathematica* can manipulate data structures in their entirety, rather than in a piecemeal fashion.

The (APL-like) ability to work with all of the elements of a matrix at once (e.g., shifting the positions of the matrix elements or performing arithmetic operations on the corresponding elements of several matrices) allows us to write CA programs that are quite concise, yet very readable.

- *Mathematica* has many functional programming constructs.

Programs can be written in the functional style of programming, which is closer to the way a person might think about solving a computing problem than to the way a computer would perform the computation. This high-level programming style enables us to focus on the details of the CA model we are developing, rather than on the details of the program that implements the model.

- *Mathematica* is an interpreted language.

The ability to create and debug, in a piecemeal fashion, the code fragments that make up a CA program, greatly decreases the time needed for a programmer to create a working program, which is a more valuable commodity (at least to us) than the time taken by the computer to run the program.

- *Mathematica* has extensive, easy-to-use graphics capabilities.

CA programs often produce patterns, both of behavior and of structure, which can be most easily detected by visualization. Producing and working with graphics in *Mathematica* is very simple and allows many different types of graphical representation to be tried out to find the most informative graphical format to use for a particular CA model.

- *Mathematica* has extensive numeric capabilities.

It is very useful for classification purposes, to be able to assign various quantitative measures (i.e., numerical values) to CA results and this is easy to do using the built-in functions and packages for statistical analysis that are provided by *Mathematica*.

- *Mathematica* uses an integrated computing environment.

CA programs can be developed, run, visualized, analyzed, and even written up in a single document, known as a notebook, which is very convenient.

Based on all of these factors, we have found that for a machine with a reasonably fast CPU and sufficient RAM and/or virtual memory, *Mathematica* is a nearly ideal computing environment for carrying out CA modeling investigations.

How Should this Book Be Used?

The material in this book (which was extensively field-tested in a computer simulation course, Materials Science and Engineering 382 at the University of Illinois at Urbana-Champaign) is specifically designed for use both inside and outside the classroom by any undergraduate or graduate student, academic or industrial researcher, or amateur science enthusiast who is interested in using a computer to study some physical, biological, chemical, or social phenomenon involving interacting entities.

To make this book useful to the members of this diverse audience, it has the following features:

- A large number of CA models currently being used to study apparently disparate physical, chemical, biological, and social systems are discussed. This is done to show you the usefulness of CA modeling and in the hope that you will come across one or more subjects that you find interesting enough to want to study further, using your own CA modeling ideas.

Note: Our approach to presenting these CA models is complementary to the approach used in research computer simulation articles, where a great deal of space is given to the discussion of background and to the analysis of results, but algorithmic details are rarely discussed and programming code is almost never shown. As a reviewer of the manuscript of this book perceptively noted, "This is not a textbook, nor is it a monograph. It is a clever-tool book or a clever tool-book." Accordingly, while we do mention the science underlying a particular CA and we often present a result obtained from running the CA, our primary focus is on explaining thoroughly the construction of a *Mathematica* program for the CA (if you are interested in pursuing that subject in more depth, the references cited at the end of the chapter will give you additional background and discussion of results, and they will also provide you with additional references on the topic).

- The chapters are written so that you can read them in any order, based on your own personal interests, with three exceptions:

If you are not familiar with the *Mathematica* programming language, you should look at the self-contained tutorial in Appendix A before reading the chapters, in order to familiarize yourself with how the language works.

Note: The tutorial does not attempt the herculean task of explaining all of the many aspects of the *Mathematica* system. Nor does it define all of the built-in functions used in the CA programs in the book (Appendix B does illustrate the use of many of the built-in functions for manipulating lists). You will find *Mathematica*'s on-line help to be very useful when you are working with *Mathematica*'s built-in functions. It will also be helpful to have a copy of the *Mathematica* book by Wolfram to see simple illustrations of the use of these functions.

The basic CA programming tools used throughout the book are laid out in the first chapter. Reading this chapter before the other chapters is mandatory, as it will make the CA programs in the book easier to understand and it will help you in writing your own CA programs.

CAs dealing with diffusive processes build upon the programs developed in the Random Walkers chapter which should therefore be read before the chapters that follow it.

- The introductory section of each chapter indicates apparently unrelated phenomena in different fields of science that can be modeled using the CA that is developed in the chapter.

Note: We have found that seeing connections between the behaviors of apparently disparate systems helps us to bring a broader, enriched perspective to whatever particular system we are looking at (and to escape the tunnel vision that unfortunately typifies a great deal of research).

- The diskette that accompanies the book contains notebooks with all of the CA programs used in the book. These can be downloaded to your computer and you can use them as starting points for your own excursions in CA modeling. The diskette is formatted in a DOS-compatible format and should be readable by any IBM-compatible computer, Macintosh computer (system 7.5 or later), NeXT computer, or UNIX machine. The files are available in several operating system formats on the diskette: DOS, Macintosh, and UNIX archives. The diskette contains a README.TXT file with further instructions on its usage.

In addition to the diskette, these materials are available via the World-Wide Web from the TELOS Web site. To access these materials, point your Web browser at http://www.telospub.com.

A Final Thought

Writing this book was a very enjoyable experience for us because we both find cellular automata modeling to be endlessly fascinating and *Mathematica* hacking to be inherently pleasurable. We hope that you will find the book to be equally enjoyable to read and that it inspires you to launch your own CA explorations of the world around you.

Reference

Resnick, Mitchel. *Turtles, Termites, and Traffic Jams.* MIT Press, (1994).

Acknowledgments

I would like to thank Carole for accommodating my unusual circadian rhythms.

Two academic colleagues, Nigel Goldenfeld and Hamish Fraser, provided much appreciated encouragement during my transition from equations to algorithms.

Allan Wylde has been the publisher of all of my books to date and he has been essential to their success. I am especially appreciative of his support even when, at the last minute, I changed my mind about writing a different book and decided to write this book instead.

A number of individuals at TELOS/Springer Verlag worked on this project: Walter Borden, Keisha Sherbecoe, Paul Wellin, and Robert Wexler. Don DeLand and Joe Kaiping did an excellent job of formatting the text and code. I thank all of them for their efforts and for their patience in dealing with an obsessive-compulsive author.

Finally, I would also like to express my appreciation to all of the participants of the O.J. Simpson trial. The television coverage of the trial served as a backdrop throughout my work on this book and whenever I worried that a CA simulation was behaving unrealistically, a glance from my computer to my television reassured me that far more surrealistic things happen in the real world than in my computer-generated microworlds.

—Richard J. Gaylord

I would like to acknowledge the continual encouragement of Professor Mamoru Baba, Professor Yasubei Kashiwaba, Kouji Oota, and Shinji Kikuchi during the writing of this book.

Thanks are due to Professors Kiyoshi Nishikawa, Masahiko Suhara of Kanazawa University, and Tsutomu Sato of Hirosaki University, and Masayuki Suzuki at Iwate University for their valuable advice.

I also would like to thank my parents, Shigeo and Etsuko Nishidate, my aunts Kuniko Kudou and Noriko Okazaki, and my uncle Takakazu Seki, and my father's golden retriever Bess for their support and encouragement.

—Kazume Nishidate

Contents

13   Ant Colony Activity   131

14   Predator-Prey Ecosystems   143

15   Contagion in Excitable Media   155

16   The Evolution of Cooperation and the Spatial Prisoner's
      Dilemma Game   173

      Appendix A: *Mathematica* Programming Tutorial   185

      Appendix B: Working with Lists   223

      Appendix C: Program Listing   231

      Index   257

1 | A Toolkit for Programming Cellular Automata

Introduction

The structure of a CA is extremely simple when written in the *Mathematica* programming language: a matrix is set up, a function (CA rule) is defined, and the function is repeatedly applied to the matrix. In this chapter we will lay the foundations for the CA programming in the chapters that follow. We first give both a physical and a computational definition of a cellular automaton. Then we discuss the operation of code fragments that repeatedly appear in CA programs throughout the book. Reading the programs in the book with an understanding of how these code fragments work will allow you to focus on the principal programming task in writing a CA program in *Mathematica*—defining the CA rules.

A Scientific Definition of a Cellular Automaton

Scientists generally define a cellular automaton (CA) to be a *discrete dynamical system*, where space, time, and the states of the system are all discrete and have the following properties:

- Space is represented by a regular lattice in one, two, or three dimensions.
- Each site, or *cell*, in (or on) the CA lattice can be in one of a finite number of states. The states are represented by integer number values.

Note: A variant of the CA, known as a coupled map-lattice (CML) or cell-dynamic scheme (CDS), uses continuous lattice site values. In this book, lattice site values can be integers, reals, or symbols, as well as tuplets (lists) with elements of any type, where the choice of value type is determined by the "physics" of the system being modeled (e.g., when objects in the system have more than one changeable attribute or property, the values of these quantities are grouped in an n-tuple).

- The CA system *evolves* over a succession of *time steps*. The values of all of the sites in the lattice are *updated* synchronously in each time step.

Note: While the values of all of the sites are updated in each time step, the value of a site need not change in a given time step.

- Site values are updated using a set of rules (known as a *lookup table*) that take the values of the site and its neighboring sites into account.

A Computational Definition of a Cellular Automaton

Since a CA model is always expressed as an algorithm and is invariably implemented as a computer program and run on a computer, it's useful to give a computational definition of a CA.

A CA is a computer program in which the following computations are performed, in order:

- A matrix is created with specific element values (integer, real, symbol, or list).
- A function, or a set of functions, that can be used to change the value of a matrix element, based on the values of the element and nearby elements, is defined.
- The function is applied (repeatedly) to the matrix, each time changing the values of all of the matrix elements simultaneously.

The CA Lattice

We will be working with two-dimensional cellular automata, employing a rectangular lattice. A simple square lattice is shown below.

```
Table["+", {5}, {5}] // MatrixForm
```

```
+   +   +   +   +
+   +   +   +   +
+   +   +   +   +
+   +   +   +   +
+   +   +   +   +
```

Note: The use of a rectangular lattice in our computations does not limit us to CAs based on this type of lattice. CAs based on other two-dimensional lattice systems (such as trigonal and hexagonal lattices) can also be handled using the rectangular lattice.

Each lattice site has a value. As an example, we show below a simple random Boolean lattice where lattice sites have values of 0 or 1.

```
(randomBoolean = Table[Random[Integer], {5}, {5}])//MatrixForm
```

```
0   0   1   0   1
1   1   0   0   0
0   1   0   0   0
1   0   1   1   0
0   1   1   0   0
```

Note: While many CAs use 0's and 1's, some CAs use other integer-valued numbers, real-valued numbers, symbols, lists, or a mixture of value types.

A graphical representation of a CA is especially useful for observing patterns of structure and/or behavior that occur during the evolution of the CA. One way to visualize a CA lattice *configuration* is by using a RasterArray graphics where each lattice site value is transformed into a colored rectangle.

Below, the RasterArray representation of the above randomBoolean CA lattice is shown.

```
Show[Graphics[RasterArray[Reverse[randomBoolean] /.
              {1 -> RGBColor[1, 0, 0], 0 -> RGBColor[0, 1, 0]}],
        AspectRatio -> Automatic,
        GridLines -> {Range[0, 5], Range[0, 5]}] ]
```

-Graphics-

Note: The `Reverse` of the `randomBoolean` lattice and `GridLines` are both used to create the figure above in order to make it easier to see the correspondence between the sites in the `randomBoolean` matrix and the `RasterArray` cells in the figure.

The Neighborhood

In updating the values of a site in a CA lattice, it is necessary to consider the site's value and the values of the sites in its vicinity. The set of sites that is employed in a CA program depends on the specific "physics" of the system being modeled. Here we show some of the CA neighborhoods that are commonly employed in this book.

The two most commonly used CA neighborhoods are shown below:

```
            N
            |
        W - X - E
            |
            S
```

The von Neumann Neighborhood

The set of five sites consisting of the X site and its four nearest neighbor sites lying north (above), east (right), south (below), and west (left) of X are known as the *von Neumann* neighborhood of X.

```
       NW   N   NE
         \  |  /
       W  - X -  E
         /  |  \
       SW   S   SE
```

The Moore Neighborhood

The set of nine sites consisting of the X site, the four nearest neighbor sites in its von Neumann neighborhood, and the four nearest neighbor sites lying northeast, southeast, southwest, and northwest of X are known as the *Moore* neighborhood of X.

Another neighborhood that we will use often in this book (primarily to model diffusive processes) is shown below:

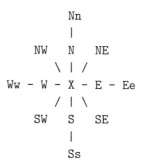

The MvonN Neighborhood

This neighborhood consists of the nine sites in the Moore neighborhood of X, and the four next nearest neighbors lying in the north, east, south, and west direction. For lack of a better name, we'll refer to this as the MvonN neighborhood of X.

Determining the neighbors of a site in the CA lattice is straightforward. The Moore neighborhoods of the sites on a small rectangular lattice are illustrated below.

```
(Lat = Partition[Range[25], 5]) // MatrixForm
```

```
 1    2    3    4    5
 6    7    8    9    10
11   12   13   14   15
16   17   18   19   20
21   22   23   24   25
```

```
Wrap = Join[{Last[#]}, #, {First[#]}]&;
(LatNeighborhoods =
       Partition[Wrap[Map[Wrap, Lat]], {3, 3}, {1, 1}]) // MatrixForm
```

```
25 21 22    21 22 23    22 23 24    23 24 25    24 25 21
 5  1  2     1  2  3     2  3  4     3  4  5     4  5  1
10  6  7     6  7  8     7  8  9     8  9 10     9 10  6

 5  1  2     1  2  3     2  3  4     3  4  5     4  5  1
10  6  7     6  7  8     7  8  9     8  9 10     9 10  6
15 11 12    11 12 13    12 13 14    13 14 15    14 15 11

10  6  7     6  7  8     7  8  9     8  9 10     9 10  6
15 11 12    11 12 13    12 13 14    13 14 15    14 15 11
20 16 17    16 17 18    17 18 19    18 19 20    19 20 16
```

```
15 11 12     11 12 13     12 13 14     13 14 15     14 15 11
20 16 17     16 17 18     17 18 19     18 19 20     19 20 16
25 21 22     21 22 23     22 23 24     23 24 25     24 25 21


20 16 17     16 17 18     17 18 19     18 19 20     19 20 16
25 21 22     21 22 23     22 23 24     23 24 25     24 25 21
 5  1  2      1  2  3      2  3  4      3  4  5      4  5  1
```

Looking at the outermost neighborhoods in LatNeighborhoods, the nearest neighbors of sites lying at various locations along the borders of Lat can be identified as follows:

- the nearest neighbor west of a site on the left border is the site in the same row on the right border.
- the nearest neighbor northwest of a site on the left border is the site in the previous row on the right border.
- the nearest neighbor southwest of a site on the left border is the site in the following row on the right border.
- the nearest neighbor east of a site on the right border is the site in the same row on the left border.
- the nearest neighbor northeast of a site on the right border is the site in the previous row on the left border.
- the nearest neighbor southeast of a site on the right border is the site in the following row on the left border.
- the nearest neighbor north of a site on the top border is the site in the same column on the bottom border.
- the nearest neighbor northwest of a site on the top border is the site in the previous column on the bottom border.
- the nearest neighbor northeast of a site on the top border is the site in the following column on the bottom border.
- the nearest neighbor below a site on the bottom border is the site in the same column on the top border.
- the nearest neighbor southwest of a site on the bottom border is the site in the previous column on the top border.
- the nearest neighbor southeast of a site on the bottom border is the site in the following column on the top border.
- the nearest neighbor northwest of the site in the top left corner is the site in the bottom right corner.
- the nearest neighbor northeast of the site in the top right corner is the site in the bottom left corner.

- the nearest neighbor southwest of the site in the bottom left corner is the site in the top right corner.

- the nearest neighbor southeast of the site in the bottom right corner is the site in the top left corner.

Dealing with Boundaries

The boundary conditions that are used in a CA program depend on the specific "physics" of the system being modeled. Traditionally, the lattice sites in the neighborhood of a border site are different for different boundary conditions. For example, under periodic boundary conditions, border sites have sites on the opposing border(s) as neighbors (e.g., a site on the left border has the site in the same row on the right border as its left neighbor) while border sites under absorbing boundary conditions have no off-lattice neighbors (e.g., a site on the left border has no left neighbor).

In our computations, lattice sites in the neighborhood of a border site are computed in the same way (which was described in the previous section) under all boundary conditions and different boundary conditions are dealt with in the manner described below.

Periodic Boundaries

This boundary, commonly used in CA models of *infinite* systems, is the default boundary condition.

Note: In a CA model of mobile objects moving in space, the wrap-around boundary condition results in objects leaving the system at one border location and reentering at the opposing border location.

Absorbing (Open) and Reflecting (Closed) Boundaries and Sinks and Sources

These boundary conditions are implemented by using a unique value for border sites, different from the values used for interior sites, and employing rules that specify what happens to the value of a site in the neighborhood of a border site and that leave the value of a border site unchanged.

Note: By using unique values for border sites, rules can be written for a CA model of mobile objects moving in space that take an object occupying a site adjacent to a border site and freeze the object in place, turn the object around, or remove the object from the system or, if the site is empty, put an object on the site.

Moving Boundaries

One way to avoid having to deal with specific boundary conditions is to adjust the size of the CA lattice on each time step so that nothing *interesting* happens on the border sites.

Writing CA Rules

Rules that update the value of a lattice site, based on the values of the sites in the neighborhood of the site, have one of the following general forms, depending on the type of neighborhood:

Rules that are applied to a site and the nearest neighbor sites in its von Neumann neighborhood have the form

$$update[site, N, E, S, W]$$

where the five arguments represent the value of the site and the values of the four nearest neighbors in the N, E, S, W directions.

Rules that are applied to a site and the nearest neighbor sites in its Moore neighborhood have the form

$$update[site, N, E, S, W, NE, SE, SW, NW]$$

where the nine arguments represent the value of the site, the values of the four nearest neighbors in the N, E, S, W directions, and the values of the four nearest neighbors in the NE, SE, SW, NW directions.

Rules that are applied to a site and the nearest neighbor sites in its Moore neighborhood and the next nearest neighbors in its von Neumann neighborhood have the form

$$update[site, N, E, S, W, NE, SE, SW, NW, Nn, Ee, Ss, Ww]$$

where the thirteen arguments represent the value of the site, the values of the four nearest neighbors in the N, E, S, W directions, the values of the four nearest neighbors in the NE, SE, SW, NW directions, and the values of the four next nearest neighbors in the N, E, S, W directions.

There is no single formula for creating CA update rules. The rules in some CAs may depend on the values of the neighborhood sites while the rules in other CAs may depend on some quantity that is computed from the values of the neighborhood sites. Additionally, some CAs may use two or more sets of rules (which need not depend on the same set of neighborhood sites) that are applied consecutively. One of the primary purposes of this book is to show a wide sampling of CA rule writing so users can develop some insight into how to approach rule making for their own CA programs.

Applying CA Rules

The following functions can be used to apply rules (which we'll call update here) to the sites of a CA lattice.

For a rule that depends on the values of sites in the von Neumann neighborhood of a site, we can use:

```
VonNeumann[update, #]&
```

where

```
VonNeumann[func__, lat_] :=
    MapThread[func, Map[RotateRight[lat, #]&,
            {{0, 0}, {1, 0}, {0, -1}, {-1, 0}, {0, 1}}], 2]
```

For a rule that depends on the values of sites in the Moore neighborhood of a site, we can use:

```
Moore[update, #]&
```

where

```
Moore[func__, lat_] :=
  MapThread[func, Map[RotateRight[lat, #]&,
            {{0, 0}, {1, 0}, {0, -1}, {-1, 0}, {0, 1},
             {1, -1}, {-1, -1}, {-1, 1}, {1, 1}}], 2]
```

For a rule that depends on the values of sites in the MvonN neighborhood of a site, we can use:

```
MvonN[update, #]&
```

where

```
MvonN[func__, lat_] :=
  MapThread[func, Map[RotateRight[lat, #]&,
            {{0, 0}, {1, 0}, {0, -1}, {-1, 0}, {0, 1},
             {1, -1}, {-1, -1}, {-1, 1}, {1, 1},
             {2, 0}, {0, -2}, {-2, 0}, {0, 2}}], 2]
```

The effect of applying these functions to a CA lattice can be seen using a simple 3-by-3 matrix:

```
(mat3 = Partition[Range[9], 3]) // MatrixForm
```

```
1   2   3
4   5   6
7   8   9
```

Applying an undefined function g to mat3, with the VonNeumann function gives

```
VonNeumann[g, #]&[mat]
```

```
{{g[1, 7, 2, 4, 3], g[2, 8, 3, 5, 1], g[3, 9, 1, 6, 2]},

 {g[4, 1, 5, 7, 6], g[5, 2, 6, 8, 4], g[6, 3, 4, 9, 5]},

 {g[7, 4, 8, 1, 9], g[8, 5, 9, 2, 7], g[9, 6, 7, 3, 8]}}
```

The result is a 3-by-3 matrix in which each element is a function call to five arguments. To identify the arguments in these function calls, we can focus on the center site in the following 5-by-5 matrix.

```
(mat5 = Partition[Range[25], 5] /. {13 -> site, 8 ->N, 14 -> E,
           18 -> S, 12 -> W, 9 -> NE,  19 -> SE, 17 -> SW, 7 -> NW,
                3 -> Nn, 15 -> Ee, 23 ->Ss, 11 -> Ww}) // MatrixForm
```

```
1      2      Nn     4      5
6      NW     N      NE     10
Ww     W      site   E      Ee
16     SW     S      SE     20
21     22     Ss     24     25
```

Using mat5 with each of the three anonymous functions defined above and the undefined function g, the values of the center site in the resulting matrix are

```
VonNeumann[g, #]&[mat5][[3, 3]]
```

```
g[site, N, E, S, W]
```

```
Moore[g, #]& [mat5][[3, 3]]
```

```
g[site, N, E, S, W, NE, SE, SW, NW]
```

```
MvonN[g, #]&[mat5][[3, 3]]
```

```
g[site, N, E, S, W, NE, SE, SW, NW, Nn, Ee, Ss, Ww]
```

Overall, applying an update function to a CA lattice, using one of the three anonymous functions defined above, results in a lattice in which a *new* value has been calculated for each CA lattice site by applying the update function to the values of the sites in the neighborhood of the corresponding site in the original lattice.

Note: While the update function appears to be applied simultaneously to each of the lattice site neighborhoods, *Mathematica* evaluates these function calls in a serial, typewriter fashion (see the Tutorial in the Appendix for a discussion of the *Mathematica* evaluation mechanism). This allows us to attach a counter to a rule definition if we wish, in order to determine how many times the rule is used during the evolution of the CA.

CA Evolution

The CA can be made to evolve over a specified number of time steps, by repeatedly applying one of the anonymous functions given above to the CA lattice that number of times. For example,

```
NestList[VonNeumann[update, #]&, initConfig, t]
```

If it is expected the CA may, at some point, stop evolving permanently (as opposed to a temporary pause), we can use the FixedPoint function with three argument values to halt the evolution of the CA when the system becomes invariant. For example,

```
FixedPointList[Moore[update, #]&, initConfig, t]
```

If we want to stop the CA when a certain condition has been met, the system becomes static, or a specified number of time steps have taken place (the progam will stop running as soon as any of these situations occurs), we can use the FixedPoint operation with four argument values, the last of which is the Same Test option (see the Tutorial in the Appendix for a discussion of this option). For example,

```
FixedPointList[MvonN[update, #]&,  initConfig, t, SameTest -> ...]
```

Note: If we are only interested in the final configuration of the CA lattice, and not the intermediate configurations, we can use the Nest and FixedPoint functions.

Watching the CA

Watching an animation of an evolving CA (rather than viewing a series of snapshots) is helpful for identifying patterns that occur while the CA is running.

We can create a "flip-card" animation of the evolution of a CA using the following command:

```
ShowCA[list_, opts___]:=
 Map[Show[Graphics[RasterArray[list[[#]] /.
   {val1 -> RGBColor[x1, y1, z1],
    val2 -> RGBColor[x2, y2, z2], ...}]], opts]&, Range[Length[list]]]
```

2 | The Game of Life

Introduction

The Game of Life is the exemplar of a cellular automaton (CA) and hence serves as a good starting point for our work with cellular automaton programming. Originally developed by John Conway, a British mathematician, the Game of Life was supposedly the first program run on a parallel processing computer. In fact, it has been estimated that more computer time has been spent running the Game of Life program than any other computer program. While the Game of Life is an abstract "toy" system that has not (yet) been found to directly represent any specific natural system, it has been the springboard for the study of so-called "artificial life" systems because of the amazingly complex behaviors displayed by some of the patterns that occur during the running of the CA. We will implement the Game of Life CA in the fastest high-level way we know of, by generating and using a lookup table consisting of 512 rules for updating sites on the Game of Life lattice.

The Game of Life CA

The System

The CA is *played* on an *n*-by-*n* square Boolean lattice, where each lattice site has a value of either 0 or 1. The boundaries of the lattice are periodic. A lattice site whose value is 1 is said to be alive and a lattice site whose value is 0 is said to be dead. The initial lattice configuration for a random distribution of dead and living sites on the lattice, is given by

```
initConfig = Table[Random[Integer], {n}, {n}]
```

In the discussion of the Game of Life program that follows, we will demonstrate how the various quantities in the program work, using the small 4-by-4 lattice, gameBoard.

```
(gameBoard =
  {{0, 0, 0, 0}, {1, 0, 1, 0}, {0, 0, 0, 0}, {1, 0, 0, 1}})//MatrixForm
```

```
0    0    0    0
1    0    1    0
0    0    0    0
1    0    0    1
```

The Moore neighborhoods of the sites in gameBoard, gameBoardNbrhds, are shown below:

```
Wrap = Join[{Last[#]}, #, {First[#]}]&;

(gameBoardNbrhds =
  Partition[Wrap[Map[Wrap, gameBoard]], {3, 3}, {1, 1}]) // MatrixForm
```

```
1 1 0    1 0 0    0 0 1    0 1 1
0 0 0    0 0 0    0 0 0    0 0 0
0 1 0    1 0 1    0 1 0    1 0 1

0 0 0    0 0 0    0 0 0    0 0 0
0 1 0    1 0 1    0 1 0    1 0 1
0 0 0    0 0 0    0 0 0    0 0 0

0 1 0    1 0 1    0 1 0    1 0 1
0 0 0    0 0 0    0 0 0    0 0 0
1 1 0    1 0 0    0 0 1    0 1 1

0 0 0    0 0 0    0 0 0    0 0 0
1 1 0    1 0 0    0 0 1    0 1 1
0 0 0    0 0 0    0 0 0    0 0 0
```

The CA Rules

In each time step, the value of each site on the CA lattice is updated according to the value of the site and the number of living nearest neighbor sites in its Moore neighborhood (rules of this sort are called semi-totalistic). The *life and death* rules for updating a lattice site are as follows:

Life and Death Rules

- Any site (dead or alive) with three living nearest neighbor sites stays alive or is born.

- A living site with two living nearest neighbor sites remains alive.
- All other sites either remain dead or die.

Traditionally, these life and death rules would be translated directly from words into code (a program that does this will be shown later). However, there is a much more efficient (in terms of the run-time speed of the resulting program) way of expressing these rules.

We can specify each of the 2^9 possible Moore neighborhood configurations (a configuration is a specification of the values of all nine sites in the neighborhood), and write a life and death update rule specific to that configuration. However, typing in 512 update rules requires too much work (especially since we type so poorly), so instead we will have a lookup table of these rules generated for us.

Creating the Game of Life Lookup Rule Table

The set of all possible Moore neighborhood configurations is computed using

```
LiveConfigs =
    Join[Map[Join[{0}, #]&, Permutations[{1, 1, 1, 0, 0, 0, 0, 0}]],
        Map[Join[{1}, #]&, Permutations[{1, 1, 1, 0, 0, 0, 0, 0}]],
        Map[Join[{1}, #]&, Permutations[{1, 1, 0, 0, 0, 0, 0, 0}]]];

DieConfigs =
    Complement[Flatten[
        Map[Permutations, Map[Join[Table[1, {#}], Table[0, {(9 - #)}]]&,
            Range[0, 9]]], 1], LiveConfigs];
```

LiveConfigs and DieConfigs are nested lists. LiveConfigs contains lists of those Moore neighborhood configurations whose center site value will be updated to a value of 1, and DieConfigs contains lists of those Moore neighborhood configurations whose center site value will be updated to a value of 0 (in all of these lists, the center site value is given by the first element in the list). We can calculate the number of life and death configurations, respectively.

```
Map[Length, {LiveConfigs, DieConfigs}]
```

```
{140, 372}
```

We now create the update rules for each of the LiveConfigs and DieConfigs configurations as follows:

```
Apply[(update[##] = 1)&, LiveConfigs, 1];

Apply[(update[##] = 0)&, DieConfigs, 1];
```

We show how this works below for a set of nine DieConfigs configuration.

```
Apply[(update[##] = 0)&,
      Permutations[{1, 1, 1, 1, 1, 1, 1, 1, 0}], 1];

?update
```

Global`update

update[0, 1, 1, 1, 1, 1, 1, 1, 1] = 0

update[1, 0, 1, 1, 1, 1, 1, 1, 1] = 0

update[1, 1, 0, 1, 1, 1, 1, 1, 1] = 0

update[1, 1, 1, 0, 1, 1, 1, 1, 1] = 0

update[1, 1, 1, 1, 0, 1, 1, 1, 1] = 0

update[1, 1, 1, 1, 1, 0, 1, 1, 1] = 0

update[1, 1, 1, 1, 1, 1, 0, 1, 1] = 0

update[1, 1, 1, 1, 1, 1, 1, 0, 1] = 0

update[1, 1, 1, 1, 1, 1, 1, 1, 0] = 0

It would take up too much space to print out (and be too boring to read) all 512 update rules, but we can check that we have, in fact, generated all of the update rules.

```
Length[DownValues[update]]
```

512

Applying the CA Rules

The update rules are applied to the CA lattice using the anonymous function

```
Moore[update, #]&
```

where

```
Moore[func__, lat_] :=
  MapThread[func, Map[RotateRight[lat, #]&,
            {{0, 0}, {1, 0}, {0, -1}, {-1, 0}, {0, 1},
             {1, -1}, {-1, -1}, {-1, 1}, {1, 1}}], 2]
```

The Moore function (as we discussed in the Toolkit chapter) applies to each site in the `lat` lattice a function `func`, whose arguments are the values of the nine sites in the Moore neighborhood of the site.

To see how this anonymous function works with the life and death rules, we can apply it to gameBoard:

```
Moore[update, #]&[gameBoard]//MatrixForm
```

```
1  1  0  0
0  0  0  0
1  1  0  0
0  0  0  0
```

The CA lattice is updated until the result ceases to change or for a specified number of times, t, using the `FixedPoint` function

```
FixedPointList[Moore[update, #]&, initConfig, t]
```

We can see how this function works by updating gameBoard over 10 time steps.

```
{FixedPointList[Moore[update, #]&, gameBoard, 10]} // MatrixForm
```

```
0 0 0 0    1 1 0 0    0 0 0 0    0 0 0 0
1 0 1 0    0 0 0 0    0 0 0 0    0 0 0 0
0 0 0 0    1 1 0 0    0 0 0 0    0 0 0 0
1 0 0 1    0 0 0 0    0 0 0 0    0 0 0 0
```

Note: Only three time steps were executed above because the lattice stopped changing after all of the site values became zero when the second time step was executed.

The overall program can now be constructed from these code fragments.

The Program

```
OblaDeOblaDa[n_, t_]:=
Module[{initConfig, Moore, update, LiveConfigs, DieConfigs},

 initConfig = Table[Random[Integer],{n},{n}] ;

 LiveConfigs =
   Join[Map[Join[{0}, #]&, Permutations[{1, 1, 1, 0, 0, 0, 0, 0}]],
        Map[Join[{1}, #]&, Permutations[{1, 1, 1, 0, 0, 0, 0, 0}]],
        Map[Join[{1}, #]&, Permutations[{1, 1, 0, 0, 0, 0, 0, 0}]]];

 DieConfigs =
   Complement[Flatten[
     Map[Permutations, Map[Join[Table[1, {#}], Table[0, {(9 - #)}]]]&,
         Range[0, 9]]], 1], LiveConfigs];

 Apply[(update[##] = 1)&, LiveConfigs, 1];
 Apply[(update[##] = 0)&, DieConfigs, 1];

 Moore[func__, lat_] :=
  MapThread[func, Map[RotateRight[lat, #]&,
           {{0, 0}, {1, 0}, {0, -1}, {-1, 0}, {0, 1},
            {1, -1}, {-1, -1}, {-1, 1}, {1, 1}}], 2];

 FixedPointList[Moore[update, #]&, initConfig, t]
 ]
```

Efficiency of the Program

We can compare the efficiency of the OblaDeOblaDa progam relative to
that of the following Game of Life program, which directly implements
the three life and death rules stated earlier (at the cost of having to
determine the number of living sites in every neighborhood in each time
step).

```
LifeGoesOn[n_, t_]:=
Module[{initConfig, Moore, update},

 initConfig = Table[Random[Integer],{n},{n}] ;

 update[1, 3] := 1;
 update[0, 3] := 1;
 update[1, 4] := 1;
 update[_, _] := 0;
```

```
Moore[func__, lat_] :=
 MapThread[func, Map[RotateRight[lat, #]&,
            {{0, 0}, {1, 0}, {0, -1}, {-1, 0}, {0, 1},
             {1, -1}, {-1, -1}, {-1, 1}, {1, 1}}], 2]

FixedPointList[MapThread[update, {#, Moore[Plus, #]}, 2]&,
               initConfig, t]
]
```

Note: The LifeGoesOn program is a slightly modified (to accommodate the use of the Moore function and to gain a bit more speed) version of the program given in (Gaylord and Wellin, 1995).

The Timings below (performed on a PowerMac 8500/120) indicate that the OblaDeOblaDa program runs about twice as fast as the LifeGoesOn program (the speed difference increases as the size of the lattice and number of time steps increases).

```
{SeedRandom[4]; Timing[LifeGoesOn[200, 20]][[1]],
 SeedRandom[4]; Timing[OblaDeOblaDa[200, 20]][[1]]}
```

```
{111.133 Second, 55.0333 Second}
```

Graphics Display

To identify the patterns that occur when the Game of Life program runs, we can take the lattice configurations in the output produced by the Game of Life program and convert them into a graphic using the following program:

```
ShowLife[list_, opts___]:=
 Map[Show[Graphics[RasterArray[Reverse[list[[#]] /.
          {1 -> RGBColor[1, 0, 0], 0 -> RGBColor[0,0,1]}]]], opts]&,
    Range[Length[list]]]
```

An example of the graphical output of the Game of Life program is given below.

```
ShowLife[OblaDeOblaDa[25, 5]];
```

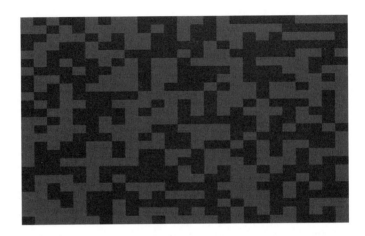

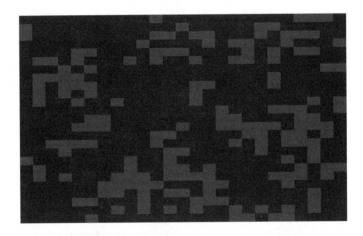

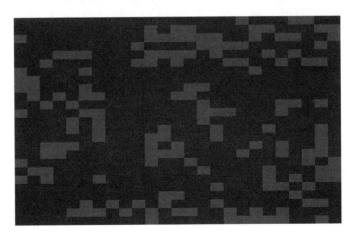

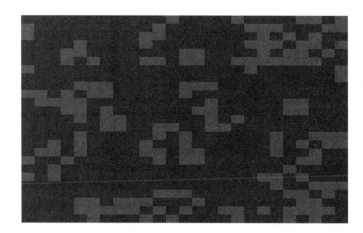

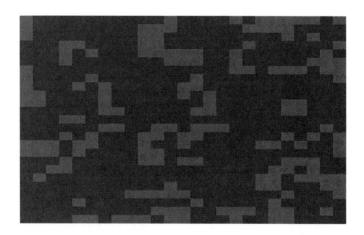

Computer Simulation Project

Persistent patterns, known as life forms, can be seen as the Game of Life evolves. Below is a catalog of some of the creatures in the Game of Life zoo.

```
glider[x_, y_] := {{x, y}, {x+1, y}, {x+2, y}, {x+2, y+1}, {x+1, y+2}}


beehive[x_, y_] :=
        {{x, y}, {x, y+1}, {x+1, y+2}, {x+1, y-1}, {x+2, y+1}, {x+2, y}}


blinker[x_, y_] :=    {{x, y}, {x+1, y}, {x+2, y}}


block[x_, y_] :=    {{x, y}, {x+1, y}, {x, y+1}, {x+1, y+1}}


barge[x_, y_] :=
      {{x, y}, {x+1, y+1}, {x+1, y-1}, {x+2, y}, {x+2, y-2}, {x+3, y-1}}


snake[x_, y_] :=
        {{x, y}, {x, y+1}, {x+1, y},   {x+2, y+1}, {x+3, y+1}, {x+3, y}}


eater[x_, y_] := {{x, y}, {x, y+1}, {x+1, y+1},
                    {x+2, y+1}, {x+2, y+3}, {x+3, y+2}, {x+3, y+3}}


beacon[x_, y_] := {{x, y}, {x+1, y}, {x, y+1}, {x+1, y+1},
                    {x+2, y+2}, {x+2, y+3}, {x+3, y+2}, {x+3, y+3}}


pentadecathlon[x_, y_] :=
        {{x, y}, {x+1, y}, {x+2, y+1}, {x+2, y-1}, {x+3, y}, {x+4, y},
        {x+5, y}, {x+6, y}, {x+7, y+1} {x+7, y-1}, {x+8, y}, {x+9, y}}


rpentomino[x_, y_] :=
                {{x, y}, {x+1, y}, {x+1, y-1}, {x+1, y+1}, {x+2, y+1}}


acorn[x_, y_] := {{x, y}, {x+1, y}, {x+1, y+2},
                        {x+3, y+1}, {x+4, y}, {x+5, y}, {x+6, y}}


toad[x_, y_] := {{x, y}, {x+1, y}, {x+1, y+1},
                        {x+2, y}, {x+2, y+1}, {x+3, y+1}}


clock[x_, y_] := {{x, y}, {x+1, y}, {x+1, y+2},
                        {x+2, y-1}, {x+2, y+1}, {x+3, y+1}}
```

Most of these patterns remain unchanged as the system evolves (these are called still-lifes), while some, such as the blinker and the clock, show periodic behavior and others, such as the glider, move across the board.

Modify the program for the Game of Life so that the lattice can be seeded with life forms. Place one or more of these life forms on the lattice and observe their behaviors over time.

References

Gaylord, Richard J. and Wellin, Paul R. *Computer Simulations with Mathematica: Explorations in Complex Physical and Biological Systems.* TELOS/Springer-Verlag (1995).

Poundstone, William. *The Recursive Universe.* Oxford University Press (1985).

Traffic Engineering

Introduction

Traffic engineering is concerned with the design of streets and highways and with the control of traffic on these roads. In an urban area, the routing of traffic needs to be considered when detours are set up so that road repairs can be made. When roadwork is done on a highway, the effect of temporary lane closures on traffic flow needs to be taken into account. That these matters are in fact quite seriously regarded by the people in the Department of Transportation is evidenced by the fact that roadwork is invariably carried out at times and locations that will maximize both inconvenience and time delay (and flagmen are stationed to ensure that no one proceeds unimpeded through a work area). Cellular automata can be used to study both routing and traffic flow on streets and highways.

Solving a Maze

One aspect of routing is determining a course by which one can go from one location to another. This is closely related to the problem of finding a path through a labyrinth. We will show how this can be done using the following simple maze on a rectangular grid.

```
labyrinth =
        {{1, 1, 1, 1, 1, 1, 1, 1, 1, 1, 1, 1, 1, 1, 1, 1, 1, 1},
         {1, 0, 0, 0, 0, 0, 1, 0, 0, 0, 0, 1, 0, 0, 0, 0, 1, 0, 1},
         {1, 1, 1, 0, 1, 0, 1, 0, 1, 1, 1, 1, 0, 1, 1, 1, 1, 0, 1},
         {e, 0, 0, 0, 1, 0, 1, 0, 0, 0, 0, 0, 0, 1, 0, 0, 1, 0, 1},
         {1, 1, 1, 0, 1, 0, 1, 0, 1, 1, 1, 1, 1, 1, 0, 1, 1, 0, 1},
         {1, 0, 0, 0, 1, 0, 1, 0, 1, 1, 0, 0, 1, 0, 0, 1, 0, 0, 1},
         {1, 0, 1, 0, 1, 0, 1, 0, 0, 0, 0, 1, 1, 1, 0, 1, 0, 1, 1},
         {1, 0, 1, 0, 1, 0, 1, 0, 1, 1, 1, 1, 0, 1, 0, 1, 0, 0, e},
         {1, 0, 1, 0, 1, 0, 0, 0, 1, 0, 0, 0, 0, 0, 0, 1, 1, 0, 1},
         {1, 0, 1, 1, 1, 1, 1, 0, 1, 1, 1, 1, 1, 1, 0, 1, 1, 0, 1},
         {1, 0, 0, 0, 0, 0, 1, 0, 0, 0, 0, 0, 0, 0, 0, 0, 0, 0, 1},
         {1, 1, 1, 1, 1, 1, 1, 1, 1, 1, 1, 1, 1, 1, 1, 1, 1, 1, 1}};
```

In the labyrinth, 1 represents a wall, 0 represents a path, and *e* indicates an entrance and an exit. The labyrinth is drawn below using red for walls, yellow for paths, and green for the entrance and exit.

```
Show[Graphics[RasterArray[Reverse[labyrinth] /.
                    {e -> RGBColor[0, 1, 0],
                     1 -> RGBColor[1, 0, 0],
                     0 -> RGBColor[1, 1, 0]}]],
       AspectRatio -> Automatic];
```

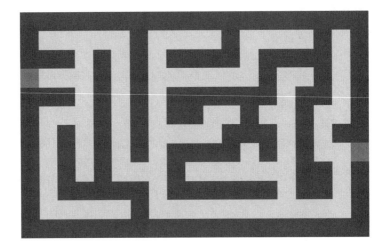

The usual procedure for solving a maze involves employing a "search" strategy while the maze is being traversed (e.g., start at the entrance and proceed along a path such that your left hand always remains in contact with a wall). While a search strategy is useful, indeed necessary, when a person is placed in a three-dimensional maze, there is a more efficient method for solving a maze map (a maze on a sheet of paper). Rather than tracing a path through the maze with a pencil (or a pen for the overconfident), it is possible to eliminate all of the cul-de-sac's (paths that culminate in a dead-end) in the maze so that all that remains is the path through the maze. This strategy can be easily implemented using a cellular automaton (Nayfeh, 1993).

Maze-Solving Cellular Automaton

The maze is a rectangular lattice consisting of sites with value 1 (a wall site), 0 (a path site), or *e* (an entance or exit site). All turns in the paths and the walls are at 90 degrees.

The CA rule for eliminating the site at the end of a dead-end path in the maze is

- A path site (a site with value 0) that has three or four nearest neighbor wall sites (sites with value 1) becomes a wall.

The CA rule for all of the other sites in the maze is

- Any other path site (including the exit and entrance) and any wall site remains unchanged.

These rules are implemented in the following six rewrite rules:

```
mazeSolve[0, 1, 1, 1, 0]   := 1

mazeSolve[0, 1, 1, 0, 1]   := 1

mazeSolve[0, 1, 0, 1, 1]   := 1

mazeSolve[0, 0, 1, 1, 1]   := 1

mazeSolve[0, 1, 1, 1, 1]   := 1

mazeSolve[x_, _, _, _, _]  := x
```

The five arguments of the mazeSolve rules are the values of a site, the north nearest neighbor to the site, the east nearest neighbor to the site, the south nearest neighbor to the site, and the west nearest neighbor to the site.

The mazeSolve rules are applied to the CA lattice using the following anonymous function (which was discussed in the Toolkit chapter):

```
VonNeumann[mazeSolve, #]&
```

where

```
VonNeumann[func__, lat_] :=
    MapThread[func, Map[RotateRight[lat, #]&,
            {{0, 0}, {1, 0}, {0, -1}, {-1, 0}, {0, 1}}], 2]
```

This anonymous function is applied repeatedly to the maze lattice using the FixedPoint function until all of the dead-end paths have been removed.

```
FixedPoint[VonNeumann[mazeSolve, #]&, maze]
```

The program for the maze-solving cellular automaton is written as

```
PathToEnlightenment[ maze_] :=
  Module[ {mazeSolve, VonNeumann},

    mazeSolve[0, 1, 1, 1, 0]    := 1;
    mazeSolve[0, 1, 1, 0, 1]    := 1;
    mazeSolve[0, 1, 0, 1, 1]    := 1;
    mazeSolve[0, 0, 1, 1, 1]    := 1;
    mazeSolve[0, 1, 1, 1, 1]    := 1;
    mazeSolve[x_, _, _, _, _]   := x;

    VonNeumann[func__, lat_] :=
      MapThread[func, Map[RotateRight[lat, #]&,
              {{0, 0}, {1, 0}, {0, -1}, {-1, 0}, {0, 1}}], 2];

    FixedPoint[VonNeumann[mazeSolve, #]&, maze]
  ]
```

We can use the PathToEnlightenment program and a maze input to
create a graphic showing side by side the maze and the path through the
maze. This is done below for the labyrinth maze.

```
Show[GraphicsArray[
 Map[Show[Graphics[RasterArray[Reverse[#] /. {e -> RGBColor[0, 1, 0],
                1 -> RGBColor[1, 0, 0], 0 -> RGBColor[1, 1, 0]}]],
          AspectRatio -> Automatic, DisplayFunction -> Identity]&,
     {#, PathToEnlightenment[#]}&[labyrinth]] ]];
```

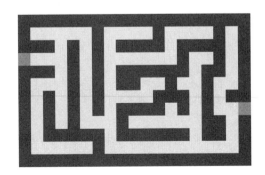

Traffic Flow

We'll look at the traffic flow on one side of a divided highway. The system consists of two lanes (with no entrances or exits along the lanes) containing cars and empty spaces (which are the same size as a car). This system is modeled in a CA using a $2 \times n$ lattice whose sites have the value 1 (indicating a space occupied by a car) or 0 (indicating an empty space). Initially, each site is occupied, with probability p, by a car (for a very long road, the density of cars on the road will approach p).

```
road = Table[Floor[p + Random[]],{2},{n}]
```

Traffic flows in the same direction along both lanes of the road (i.e., along both rows of the lattice). There is no distribution of car velocities in this simple model; at each step, a car either moves forward one space, moves sideways, or remains in place. When a car reaches one end of the road and then moves forward, it disappears from that end and reappears at the other end (this behavior corresponds to having an infinitely long road).

The rules for the behavior of a site (which is either empty or occupied by a car) during one time step in the highway CA depend on the values of the following sites:

- the site [1]
- the site ahead of the site [2]
- the site behind the site [3]
- the site adjacent to the site[4]
- the site ahead of the adjacent site [5]
- the site behind the adjacent site [6]

where the bracketed numbers above correspond to the numbering of sites in the following simple road system.

```
(roadTest= {{a, 6, 4, 5, b}, {c, 3, 1, 2, d}}) // TableForm
```

```
a   6   4   5   b
c   3   1   2   d
```

The update rules for the CA have the form

```
drive[site, site ahead, site behind, site adjacent,
            site ahead of adjacent site, site behind adjacent site]
```

where the arguments to drive are the values of the six sites listed above, in order.

Note: In the rule descriptions that follow, the sites in roadTest that correspond to the sites in the rules are indicated in square brackets.

The six rules of the road are:

- An occupied site [1] becomes empty when the site ahead of the site [2] is empty (i.e., a car that has an empty space ahead of it moves into the space).

```
drive[1, 0, _, _, _, _] := 0
```

- An occupied site [1] becomes empty when the site ahead of the site [2] is occupied and the sites adjacent to the site [4] and behind the adjacent site [6] are both empty (i.e., a car that is blocked by a car ahead of it switches lanes when there is an empty adjacent space and an empty space behind that empty adjacent space).

```
drive[1, 1, _, 0, _, 0] := 0
```

- An occupied site [1] remains occupied when the sites ahead of the site [2] and adjacent to the site [4] are both occupied (i.e., a car that is blocked by cars ahead and next to it doesn't move).

```
drive[1, 1, _, 1, _, _] := 1
```

- An empty site [1] that has an occupied site behind it [3] becomes occupied (i.e., an empty space ahead of a car becomes occupied by the car).

```
drive[0, _, 1, _, _, _] := 1
```

- An empty site [1] that has an empty space behind it [3], an occupied site adjacent to it [4], and an occupied site ahead of the adjacent site [5] becomes occupied (i.e., an empty space that is followed by an empty space and is next to a car that is blocked by a car ahead of it becomes occupied as the car switches lanes).

```
drive[0, _, 0, 1, 1, _] := 1
```

- In situations other than those given above, empty sites remain empty and occupied sites remain occupied.

```
drive[x_, _, _, _, _, _] := x
```

The following anonymous function is used to apply the drive rules to the road lattice:

```
MapThread[drive,
          {#, RotateRight[#, {0, -1}], RotateRight[#, {0, 1}],
          RotateRight[#, {1, 0}], RotateRight[#, {1, -1}],
          RotateRight[#, {1, 1}]}, 2]&
```

To illustrate how this anonymous function works, we can use the small roadTest lattice. Applying the anonymous function above to roadTest, using an undefined function *rule*, and looking at what happens to the roadTest site numbered as 1, we see that the sites that are numbered as 2 through 6 are taken, in order, as arguments to *rule*.

```
MapThread[rule,
          {#, RotateRight[#, {0, -1}], RotateRight[#, {0, 1}],
          RotateRight[#, {1, 0}], RotateRight[#, {1, -1}],
          RotateRight[#, {1, 1}]}, 2]&[roadTest][[2, 3]]
```

```
rule[1, 2, 3, 4, 5, 6]
```

The anonymous function is used with the NestList function to repeatedly apply the drive rules to the road lattice *t* times.

```
NestList[MapThread[drive,
            {#, RotateRight[#, {0, -1}], RotateRight[#, {0, 1}],
            RotateRight[#, {1, 0}],  RotateRight[#, {1, -1}],
            RotateRight[#, {1, 1}]}, 2]&, road, t]
```

We can write the program for the evolution of this CA as follows:

```
keepOnTrucking[n_, p_, t_] :=
  Module[{road, drive},

    road = Table[Floor[p + Random[]],{2},{n}];

    drive[1, 0, _, _, _, _] =  0;
    drive[1, 1, _, 0, _, 0] =  0;
    drive[1, 1, _, 1, _, _] =  1;
    drive[0, _, 1, _, _, _] =  1;
```

```
drive[0, _, 0, 1, 1, _] =  1;
drive[x_, _, _, _, _, _] := x;

NestList[MapThread[drive,
                {#, RotateRight[#, {0, -1}], RotateRight[#, {0, 1}],
                 RotateRight[#, {1, 0}],  RotateRight[#, {1, -1}],
                 RotateRight[#, {1, 1}]}, 2]&, road, t]
   ]
```

The dynamic behavior of traffic can be seen in an animation. This is shown below for 10 successive time steps.

```
Map[Show[Graphics[RasterArray[# /.
             {0 -> RGBColor[0.7, 0.7, 0.7], 1 -> RGBColor[0, 1, 0]}]],
        AspectRatio -> Automatic]&, keepOnTrucking[50, 0.7, 10]];
```

Road Obstacles

The effect of roadwork or traffic accidents on the flow of traffic in the two-lane divided highway model can be computed by making some modifications in the keepOnTrucking program.

Note the ease with which these changes can be incorporated into the model; this is one of the main advantages of a rule-based programming approach.

Using c to represent an obstacle, we can place a single obstacle at a random location along the road with

```
roadWithObstacle =
   ReplacePart[Table[Floor[p + Random[]],{2}, {n}],  c,
              {Random[Integer,{1, 2}], Random[Integer, {1, n}]}]
```

The following six rules are used to describe the movement of cars on the road when obstacles are present:

- A car that has an empty space ahead of it moves ahead.

```
driveWithObstacle[1, 0, _, _, _, _] =  0
```

- A car that is blocked by either a car or an obstacle ahead of it switches lanes when there is an adjacent empty space and an empty space or an obstacle behind that empty space.

```
driveWithObstacle[1, 1 | c, _, 0, _, 0 | c] =  0
```

- A car that is blocked by cars or obstacles ahead and next to it doesn't move.

```
driveWithObstacle[1, 1 | c, _, 1 | c, _, _] =  1
```

- An empty space ahead of a car becomes occupied by the car.

```
driveWithObstacle[0, _, 1, _, _, _] =  1
```

- An empty space that is followed by an empty space or an obstacle and is next to a car that is blocked by a car or an obstacle ahead of it becomes occupied as the car switches lanes.

```
driveWithObstacle[0, _, 0 | c, 1, 1 | c, _] =  1
```

- Obstacles remain in place and in situations other than those given above, cars don't move and empty spaces remain empty.

```
driveWithObstacle[x_, _, _, _, _, _] := x
```

The overall program for traffic flow in the presence of obstacles on the road is given by

```
indy500[n_, p_, t_] :=
  Module[{roadWithObstacle, driveWithObstacle},

    roadWithObstacle =
      ReplacePart[Table[Floor[p + Random[]],{2}, {n}],  c,
                    {Random[Integer,{1, 2}], Random[Integer, {1, n}]}];

    driveWithObstacle[1, 0, _, _, _, _] =  0;
    driveWithObstacle[1, 1 | c, _, 0, _, 0 | c] =  0;
    driveWithObstacle[1, 1 | c, _, 1 | c, _, _] =  1;
    driveWithObstacle[0, _, 1, _, _, _] =  1;
    driveWithObstacle[0, _, 0 | c, 1, 1 | c, _] =  1;
    driveWithObstacle[x_, _, _, _, _, _] := x;


    NestList[MapThread[driveWithObstacle,
                {#, RotateRight[#, {0, -1}], RotateRight[#, {0, 1}],
                 RotateRight[#, {1, 0}],  RotateRight[#, {1, -1}],
                 RotateRight[#, {1, 1}]}, 2]&, roadWithObstacle, t]
    ]
```

Computer Simulation Projects

1. Modify the keepOnTrucking program to calculate the mean velocity of the system—the number of cars moving ahead in a time step divided by the total number of cars.

   Note: This can be done by using a rule of the form

   ```
   drive[1, 0, _, _, _, _] :=  (count++; 0)
   ```

   and modifying the function used in the NestList operation in keepOnTrucking, so that at the end of each time step the value of count is placed at the end of a list whose elements are the values of count at the end of each of the preceding time steps, and the value of count is reset to 0 (see Gaylord and Wellin, 1995).

2. Use the program from the previous project to create the fundamental diagram of traffic flow—a plot of the flow or mean velocity of cars per time step versus the car density.

References

Gaylord, Richard J. and Wellin, Paul R. *Computer Simulations with Mathematica: Explorations in Complex Physical and Biological Systems*, pp. 140–3. TELOS/Springer-Verlag (1995).

Nayfeh, Basem A. 1993. "Cellular automata for solving mazes." Dr. Dobb's Journal, February, p. 32.

4 Spinoidal Decomposition and Phase-Ordering in Binary Mixtures

Introduction

Binary alloy systems undergo nonequilibrium phase separation in which domains of two stable phases grow from a thermodynamically unstable homogeneous phase as a result of quenching below a critical temperature.

Traditionally, phase-ordering phenomena have been theoretically modeled by numerically solving nonlinear partial differential equations (viz. the Ginzburg-Landau and Cahn-Hilliard equations), which amounts to space-time discretizing a continuum model. An alternative approach is to use a space-time discrete lattice from the start. This type of model is referred to as a cell dynamic scheme (CDS) (coupled maps and cellular automata are examples of CDSs).

The following three features need to be incorporated into a CDS model of spinoidal decomposition (Oono, 1987):

- The local tendency to segregate (i.e., to increase the local concentration difference between the species).
- The stability of the segregated bulk phases.
- The conservation of matter.

The order parameter, which is the concentration difference between the two components in a binary mixture, normalized between −1 and 1, can be used to describe the configuration of the binary system. The spinoidal decomposition CDS computes the evolution of the order parameter over time using the constraint that the order parameter be conserved.

In addition to using the CDS to model domain formation in spinoidal decomposition, it can be used to explain the periodic patterns formed in the microphase separation in block copolymers, which are long chain molecules (resembling an unclasped pearl necklace) consisting of n consecutive monomeric units of type A and m consecutive monomeric units of type B joined together by covalent chemical bonds (this system differs from the spinoidal system in that the covalent bonds between A and B limit the size of the domains).

We will present two programs for the deterministic CDS model of the evolution of the order parameter field for phase-ordering phenomenon: one program in which the order parameter is conserved and one program in which the order parameter is not conserved.

The Phase-Ordering CA

The Nonconserved Order Parameter Case

The model employs a square lattice of size m with periodic boundary conditions. The lattice sites have values between -1 and 1 where the value represents the order parameter.

Initially, the order parameter is randomly distributed throughout the lattice in a uniform manner between -0.1 and 0.1.

```
Table[Random[Real, {-0.1, 0.1}], {m}, {m}]
```

During each time step, the order parameter values on all of the lattice sites change. In the absense of the constraint that the order parameter be conserved, the change in the order parameter value of a site arises from (a) a diffusive effect due to the difference of the order parameter from the average order parameter of neighboring sites, and (b) a local driving force due to the chemical potential.

The order parameter values on the lattice sites are updated using the following function whose arguments are the values of a site and the eight nearest neighbors in its Moore neighborhood.

```
separate[x_, n_, e_, s_, w_, ne_, se_, sw_, nw_] :=
    1.3 Tanh[x] + d ((n + e + s + w)/6 + (ne + se + sw + nw)/12 - x)
```

The quantity `1.3 Tanh[x]` is a symmetric hyperbolic map (Oono, 1987). It has one hyperbolic source (which is an unstable fixed point corresponding to the disordered state before quenching) located at the origin and two hyperbolic sinks (which are stable fixed points corresponding to the two ordered states after quenching) symmetrically placed on either side of the unstable fixed point.

Note: The precise form of the hyperbolic map is not important for the CDS modeling.

The quantity d is a positive real number (proportional to the diffusion constant) representing the stability of bulk uniformity.

The term $(n + e + s + w)/6 + (ne + se + sw + nw)/12$ calculates an isotropic average over all of the sites in a Moore neighborhood except the center site, summing the total value of the neighbors, with sites in the N, S, E, and W directions given twice the weight of sites in the NE, SE, SW, and NW directions.

The separate rule is applied to the lattice over t time steps using

```
Nest[Moore[separate, #]&, initconfig, t] ]
```

where

```
Moore[func__, lat_] :=
  MapThread[func, Map[RotateRight[lat, #]&,
          {{0, 0}, {1, 0}, {0, -1}, {-1, 0}, {0, 1},
           {1, -1}, {-1, -1}, {-1, 1}, {1, 1}}], 2]
```

Finally, a graphics display (greyscale image) of the order parameter field is created using the ListDensityPlot function.

```
ListDensityPlot[newlat, Mesh->False, FrameTicks->None]
```

These pieces of code can be combined into a program.

The Nonconserved Order Parameter Program

```
PhaseOrderingNonConserved[d_, m_, t_]:=
Module[{initconfig, Moore, separate, newlat},

 initconfig = Table[Random[Real, {-0.1, 0.1}], {m}, {m}];

 separate[x_, n_, e_, s_, w_, ne_, se_, sw_, nw_] :=
    1.3 Tanh[x] + d ((n + e + s + w)/6 + (ne + se + sw + nw)/12 - x);

 Moore[func__, lat_] :=
    MapThread[func, Map[RotateRight[lat, #]&,
            {{0, 0}, {1, 0}, {0, -1}, {-1, 0}, {0, 1},
             {1, -1}, {-1, -1}, {-1, 1}, {1, 1}}], 2];

 newlat = Nest[Moore[separate, #]&, initconfig, t];

 ListDensityPlot[newlat, Mesh -> False, FrameTicks -> None]
 ]
```

The Conserved Order Parameter Case

The constraint that the order parameter be conserved essentially demands that when there is an exchange of order parameter values between a site and its neighbors, there is no net change of the order parameter inside the neighborhood of the site. In this case, the lattice sites are updated at each time step using (Oono, 1987)

```
[Function[y, (# + y - Moore[nnave, y])][Itn[#]]&
```

where

```
nnave[x_, n_, e_, s_, w_, ne_, se_, sw_, nw_] :=
                    (n + e + s + w)/6 + (ne + se + sw + nw)/12
```

and

```
Itn[mat_] := d (Moore[nnave, mat] - mat) + 1.3 Tanh[mat] - mat
```

The Conserved Order Parameter Program

```
PhaseOrderingConserved[d_, m_, t_] :=
Module[{initconfig nnave, Itn, newlat},

  initconfig = Table[Random[Real, {-0.1, 0.1}], {m}, {m}];

  nnave[x_, n_, e_, s_, w_, ne_, se_, sw_, nw_] :=
                    (n + e + s + w)/6 + (ne + se + sw + nw)/12;

  Moore[func__, lat_] :=
    MapThread[func, Map[RotateRight[lat, #]&,
            {{0, 0}, {1, 0}, {0, -1}, {-1, 0}, {0, 1},
            {1, -1}, {-1, -1}, {-1, 1}, {1, 1}}], 2];

  Itn[mat_] := d (Moore[nnave, mat] - mat) + 1.3 Tanh[mat] - mat;

  newlat = Nest[Function[y, (# + y - Moore[nnave, y])][Itn[#]]&,
            initconfig, t];

  ListDensityPlot[newlat, Mesh -> False, FrameTicks -> None]
  ]
```

Domain Formation

We show below some typical output from running the phase-ordering programs for the conserved order parameter constraint case and the nonconserved order parameter constraint case using a 100-by-100 lattice with d equal to 0.5, after 0, 20, and 100 time steps.

```
SeedRandom[8];
PhaseOrderingConserved[0.5, 100, 0];
```

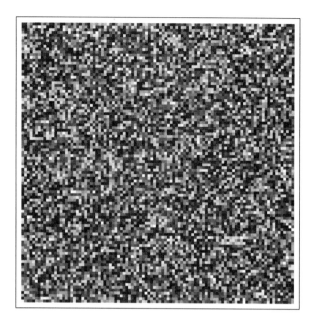

```
SeedRandom[8];
PhaseOrderingConserved[0.5, 100, 20];
```

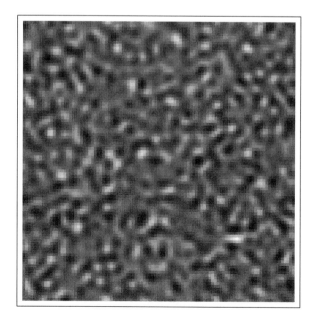

```
SeedRandom[8];
PhaseOrderingConserved[0.5, 100, 100];
```

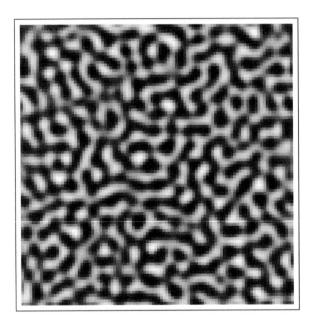

```
SeedRandom[8];
PhaseOrderingConserved[0.5, 100, 500];
```

```
SeedRandom[8]
PhaseOrderingNonConserved[0.5, 100, 0];
```

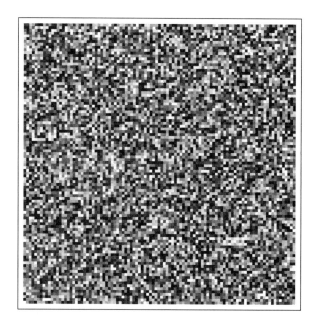

```
SeedRandom[8]
PhaseOrderingNonConserved[0.5, 100, 20];
```

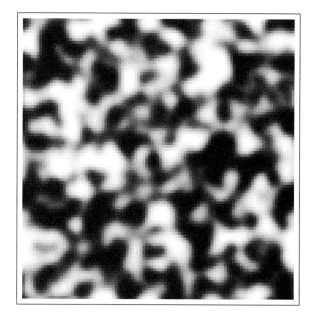

```
SeedRandom[8]
PhaseOrderingNonConserved[0.5, 100, 100];
```

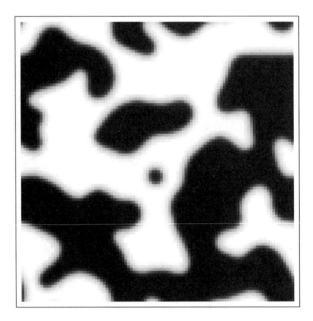

Computer Simulation Projects

1. The ListDensityPlot indicates variations in the local concentration difference between species in a monotonic fashion (i.e., as the order parameter varies from -1 to 1, the greyscale of the ListDensityPlot changes monotonically from 0 to 1). As such, it doesn't allow us to clearly see where one or the other species dominates in an interfacial region. To do this, we can assign different colors to positive and negative values of the order parameter. Use the ColorFunction option with ListDensityPlot to do this.

   Hints:

   When the ListDensityPlot is applied to the order parameter lattice, the values that vary from -1 to 1 are automatically rescaled to vary from 0 to 1. Therefore, in using the ColorFunction, the value 0.5 denotes the point at which there is a change in which species exist in greater local concentration.

```
ColorFunction ->
     (If[# < 0.5, RGBColor[2 #, 0, 0],  RGBColor[0, 2 (# - 0.5), 0]]&)
```

   where the second argument in the anonymous function results in order parameter values between -1 and 0 (which have been rescaled by the `ListDensityPlot` function to run from 0 to 0.5) being colorized between black (since $2 * 0.0 = 0$) and red (since $2 * 0.5 = 1$) and the third argument in the anonymous function results in order parameter values between 0 and 1 (which have been rescaled by the `ListDensityPlot` function to run from 0.5 to 1) being colorized between black (since $2*(0.5-0.5) = 0$) and green (since $2*(1.0-0.5) = 1$).

One problem with using the above `ColorFunction` option is that both colors shade to black at one limit. We can make the coloring range from dark red to red and from dark green to green for the two species by using

```
ColorFunction -> (If[# < 0.5,
                RGBColor[(0.2 + 0.8 (2 #)), 0, 0],
                RGBColor[0, (0.2 + 0.8 (2 (# - 0.5))), 0]]&)
```

Note: We can use `CMYKColor` or any of Mathematica's other color functions in a similar way.

2. A very simple cellular automaton, known as the Vote CA, shows the same phase-ordering phenomenon produced by the detailed CDS models given here.

The Vote CA can be described as follows:

The sites of a square lattice have values of 0 and 1. At each update, a site's value is replaced by the value possessed by the majority of the nine sites in its Moore neighborhood, with the following exceptions: if there is one more neighborhood site with a value of 1 than with a value of 0 (i.e., five sites with value 1 and four sites with value 0), the site value is updated to 0, and if there is one less neighborhood site with a value of 1 than with a value of 0 (i.e., four sites with value 1 and five sites with value 0), the site value is updated to 1. The rules of this CA are given in the table below:

```
Print["                    The CA Vote Rule Table"];
TableForm[{Range[0, 9], {0, 0, 0, 0, 1, 0, 1, 1, 1, 1}},
TableHeadings->{{"Sum over neighborhood", "New cell value"},
 None}]
```

The CA Vote Rule Table

Sum over neighborhood	0	1	2	3	4	5	6	7	8	9
New cell value	0	0	0	0	1	0	1	1	1	1

The Vote CA program can be written as

```
Vote[s_, t_]:=
 Module[{rule, init, Moore, newlat},

  init = Table[Random[Integer], {s}, {s}];

  Moore[func__, lat_] :=
    MapThread[func, Map[RotateRight[lat, #]&,
              {{0, 0}, {1, 0}, {0, -1}, {-1, 0}, {0, 1},
               {1, -1}, {-1, -1}, {-1, 1}, {1, 1}}], 2]

  rule[4] := 1;
  rule[5] := 0;
  rule[x_] := Floor[x/5];
  Attributes[rule] = Listable;

  newlat = Nest[rule[Moore[Plus, #]]&, init, t];

  ListDensityPlot[newlat, Mesh -> False, FrameTicks -> None]
 ]
```

Entering and running the Vote CA program yields

```
SeedRandom[3]
Vote[100, 0];
```

```
SeedRandom[3]
Vote[100, 50];
```

Compare the graphics produced by the PhaseOrdering and Vote programs.

Hint: To make a proper comparision of the PhaseOrdering programs with the Vote CA program, the quantity `newlat` in the `ListDensityPlot` function should be replaced with `Sign[newlat]`.

References

Oono, Yoshitsugu and Puri, S. 1987. "Computationally efficient modeling of ordering of quenched phases." Phys. Rev. Lett., 58, 836–839.

Oono, Yoshitsugu and Shiwa, Y. 1987. "Computationally efficient modeling of block copolymer and Benard pattern formations." Mod. Phys. Lett. B, 1, 49–55.

Oono, Y. and Puri S. 1988. "Study of phase-separation dynamics by use of cell dynamical systems. I. Modeling." Phys. Rev. A, 38, 434–453.

Note: This article contains the statement "Nature gives physicists phenomena, not equations." This could serve as a rallying cry of the field of algorithmic physics in which phenomena are directly modeled by algorithms or computer programs rather than by equations.

Puri, S. and Oono, Y. 1988. "Study of phase-separation dynamics by use of cell dynamical systems. II. Two-dimensional demonstrations." Phys. Rev. A, 38, 1542–1565.

5 Solidification

Introduction

Solidification is a process in which crystals form in a supercooled melt. The crystallization process itself produces some effects that tend to inhibit crystal growth: the diffusion of latent heat at the solid-liquid interface and surface tension. We present a deterministic cellular automaton model of crystal growth (Liu and Goldenfeld, 1990) that explicitly incorporates these two growth inhibition effects.

The Dendrite CA

The model employs a square lattice. Each site in the lattice has an ordered pair as a value. The first component of the ordered pair is a Boolean value, where 1 represents a site that is crystalline and 0 represents a site that is amorphous. The second component of the ordered pair is a nonnegative real number representing the temperature of the site.

Initially, the lattice is a 5-by-5 crystal seed, consisting of random $\{0, 0\}$'s and $\{1, \text{undercool}\}$, surrounded by $2m$ rows and $2m$ columns of $\{0, 0\}$'s, given by

```
seedLat =
      zerosDecorate[Table[{1, undercool} Random[Integer], {5}, {5}]]
```

where

```
zerosDecorate =
   Nest[(Prepend[Append[Map[Prepend[Append[#, {0, 0}], {0, 0}]&, #],
                Table[{0, 0}, {Length[#] + 2}]],
            Table[{0, 0}, {Length[#] + 2}]])&, #, m]&
```

The quantity $\{0, 0\}$ indicates a liquid site whose temperature is zero and the quantity $\{1, \text{undercool}\}$ indicates a crystalline site whose temperature equals the undercooling.

The crystallization process proceeds over a number of time steps, in each of which two consecutively executed events occur:

[1] An amorphous site becomes crystalline if it is adjacent to at least one crystalline site and if some condition related to the local interface curvature and temperature is met. Latent heat of crystallization is produced at a crystallizing site, which raises its temperature. An amorphous site that does not meet the prerequisite condition remains liquid and a crystalline site remains unchanged (i.e., there is no melting in the model).

The phase and temperature values of each site in the lattice are updated based on its temperature and phase and on the phases of the sites lying in its Moore neighborhood (i.e., the N, E, S, W, NE, SE, SW, and NW nearest neighbor sites).

The CA rewrite rules take nine arguments, representing the value of a lattice site and the values of the eight sites in its Moore neighborhood.

```
phase[{0, a_}, {s_, _}, {t_, _}, {u_, _},
                {v_, _}, {w_, _}, {x_, _}, {y_, _}, {z_, _}] :=
        {1, a + latheat} /; MatchQ[1, s | t | u | v] &&
            a < undercool (1 + delta Random[Real, {-1, 1}]) +
                lambda (2 (s + t + u + v) + (w + x + y + z) - 6)

phase[{r_, a_}, {_, _}, {_, _}, {_, _}, {_, _},
                {_, _}, {_, _}, {_, _}, {_, _}] := {r, a}
```

Notes:

The quantities in the first phase rule (which crystallizes a site) have the following meanings:

(a) The latent heat of crystallization is given by *latheat*.

(b) Persistent experimental noise is represented by the term *delta Random[Real, $\{-1, 1\}$]*.

(c) The noise amplitude is given by *delta*, where $0 <= \text{delta} < 1$.

(d) The capillary length is given by *lambda*.

(e) The local interface curvature is approximated by the quantity

$-$(twice the sum of the N, E, S, W site phase values

$+$ the sum of the NE, SE, SW, NW site phase values $- 6$)

The phase rule is applied to the lattice using

```
Moore[phase, #]&
```

where

```
Moore[func__, lat_] :=
  MapThread[func, Map[RotateRight[lat, #]&,
          {{0, 0}, {1, 0}, {0, -1}, {-1, 0}, {0, 1},
          {1, -1}, {-1, -1}, {-1, 1}, {1, 1}}], 2]
```

[2] Heat diffuses to adjacent sites.

The temperature of each site in the lattice is updated based on its temperature and the temperature of the nearest neighbor sites lying N, E, S, W, NE, SE, SW, and NW of the site.

The CA rewrite rules take nine arguments, representing the value of a lattice site and the values of the eight sites in its Moore neighborhood,

```
temp[{r_, a_}, {_, b_}, {_, c_}, {_, d_},
        {_, e_}, {_, f_}, {_, g_}, {_, h_}, {_, i_}] :=
        {r, a + (D/m)*((b + c + d + e)/6+ (f + g + h + i)/12 - a)}
```

where D is the diffusion constant and m is an integer number.

The temp rule is applied to the lattice using

```
Nest[Moore[temp, #]&, zerosDecorate[Moore[phase, #]], m]&
```

Note: The heat diffusion process is executed m times in each time step while the phase transformation process is only executed once in order to account for the relative speeds of the two processes. To allow the heat to diffuse m times without diffusing from one side of the lattice to the opposing side (which would be unphysical), the lattice is decorated with $2m$ rows and $2m$ columns of $\{0, 0\}$'s (using the zerosDecorate function) before the temp rule is applied.

These two processes are carried out repeatedly until the crystallization stops of its own accord or after a specified number of time steps, using

```
FixedPoint[Nest[Moore[temp, #]&, zerosDecorate[Moore[phase, #]], m]&,
        seedLat, n]
```

These pieces of code can be combined into a program.

The Program

```
dendrite[D_Real, m_, lambda_, latheat_, delta_, undercool_, n_] :=
Module[{zerosDecorate, seedLat, phase, temp, Moore},

  zerosDecorate =
    Nest[(Prepend[Append[Map[Prepend[Append[#, {0, 0}], {0, 0}]&, #],
                      Table[{0, 0}, {Length[#] + 2}]],
                Table[{0, 0}, {Length[#] + 2}]])&, #, m]&;

  seedLat =
      zerosDecorate[Table[{1, undercool} Random[Integer], {5}, {5}]];

  phase[{0, a_}, {s_, _}, {t_, _}, {u_, _},
                {v_, _}, {w_, _}, {x_, _}, {y_, _}, {z_, _}] :=
    {1, a + latheat} /; MatchQ[1, s | t | u | v] &&
              a < undercool (1 + delta Random[Real, {-1, 1}]) +
                  lambda (2 (s + t + u + v) + (w + x + y + z) - 6);

  phase[{r_, a_}, {_, _}, {_, _}, {_, _}, {_, _},
                            {_, _}, {_, _}, {_, _}, {_, _}] := {r, a};

  temp[{r_, a_}, {_, b_}, {_, c_}, {_, d_},
          {_, e_}, {_, f_}, {_, g_}, {_, h_}, {_, i_}] :=
        {r, a + (D/m)*((b + c + d + e)/6+ (f + g + h + i)/12 - a)};

  Moore[func__, lat_] :=
    MapThread[func, Map[RotateRight[lat, #]&,
              {{0, 0}, {1, 0}, {0, -1}, {-1, 0}, {0, 1},
              {1, -1}, {-1, -1}, {-1, 1}, {1, 1}}], 2];

  FixedPoint[Nest[Moore[temp, #]&, zerosDecorate[Moore[phase, #]],m]&,
            seedLat, n]
  ]
```

Growing a Crystal

In running the dendrite program, the quantities, capillary length, latent heat, diffusion constant, and *m* are first adjusted to convenient values and then the effect of different undercoolings are examined. We'll show a typical output from running the program.

```
solid = dendrite[4.0, 5, 0.015, 1, 0.1, 0.4, 40];
```

```
ListDensityPlot[
        -Map[Take[#, {160, -160}]&, Take[solid, {160, -160}]] /.
            {x_, y_?NumberQ} -> x, Mesh -> False, FrameTicks -> None]
```

-DensityGraphics-

```
ListDensityPlot[
    Map[Take[#, {160, -160}]&, Take[solid, {160, -160}]] /.
            {x_, y_?NumberQ} -> y,  Mesh->False, ColorFunction -> Hue,
                                                FrameTicks -> None]
```

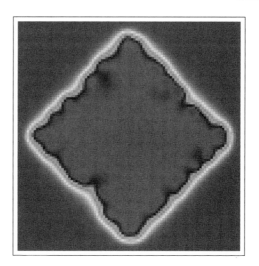

-DensityGraphics-

The black-and-white graphic shows a crystal grown at a small undercooling and the color graphic shows the temperature field for the crystal.

Note: We have cropped the pictures (removing the outer rows and columns, which have no crystalline sites) so as to enlarge the view of the crystal and its thermal field.

Computer Simulation Projects

1. Crystallization from the melt commonly involves the simultaneous growth of multiple crystals until they impinge upon one another. Modify the `dendrite` program so that there are several randomly placed crystal seeds that grow simultaneously.

2. Modify the program created in the previous project so that a graphic can be created showing the boundaries between impinging crystals.

References

Liu, Fong and Goldenfeld, Nigel. 1990. "Generic features of late-stage crystal growth." Phys. Rev. A, 42, 895–903.

Liu, Fong and Goldenfeld, Nigel. 1991. "Deterministic lattice model for diffusion-controlled crystal growth." Physica D, 47, 124–131.

Rappaz, Michel and Kurz, Wilfried. 1995. "Dendrite solidified by computer." Nature, 375 (5/11/95), 103.

6 | Snowflakes

Introduction

Crystallization is a liquid-solid phase transition in which the solid grows by adding material from the liquid environment adjacent to it. This process, which can be classified as a type of spreading phenomenon, is complicated by the fact that there is a *growth inhibition* effect whereby the crystallization of a site inhibits the crystallization of nearby sites. The molecular causes of crystal growth inhibition, including interfacial surface tension and latent heat diffusion effects, can be incorporated in a detailed lattice model (see the Solidification CA) but it is also possible to use a very simple cellular automaton model (Packard, 1986) incorporating growth inhibition in a naive fashion, which produces patterns that bear a remarkable resemblance to the observed structures of snowflakes. We will develop this cellular automaton model for the formation of snowflakes on a hexagonal lattice.

Working with a Hexagonal Lattice

We can display the centers of a small lattice of hexagons, using

```
Show[
 Graphics[{PointSize[0.06],
   Map[Point, Flatten[Table[{i Sqrt[3], j},
                             {i, 1, 10}, {j, Mod[i, 2], 7, 2}], 1]]}],
   PlotRange -> All, AspectRatio -> Automatic];
```

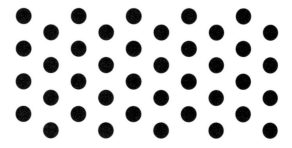

The figure illustrates that a hexagonal lattice is essentially a distorted rectangular lattice. Thus, it should be possible to model a hexagonal lattice CA using a rectangular lattice CA and converting the results onto a hexagonal lattice. To do this, the neighborhood of a site in a hexagonal lattice needs to be mapped onto the neighborhood of a site in a rectangular lattice.

Consider the following small hexagonal lattice:

```
   11        12        13        14        15        16

        21        22        BB        24        25        26

   31        AA        33        34        35        36

        41        42        43        44        45        46
```

The nearest neighbor sites of element AA in the third row are 21, 22, 31, 33, 41, 42. The nearest neighbor sites of element BB in the second row are 13, 14, 22, 24, 33, 34.

When this hexagonal lattice is converted into a rectangular lattice, these nearest neighbors sites become elements in a Moore neighborhood, which consists of the sites N, NE, E, SE, S, SW, W, NW of a site. The positioning of the nearest neighbor sites differs for sites in odd- and even-numbered rows. We can demonstrate this using our simple hexagonal lattice.

For site AA, the six nearest neighbor sites (21, 22, 31, 33, 41, 42) are positioned as follows:

```
11  12   13  14  15  16
21  22   BB  24  25  26
31  AA   33  34  35  36
41  42   43  44  45  46
```

In general, the shape of the nearest neighbor sites for a site in an odd-numbered row looks like

```
    *   *
    *       *
    *   *
```

Hence, the six nearest neighbors of a site in an odd-numbered row on a hexagonal lattice become the nearest neighbors of the site in the N, E, S, SW, W, NW directions on a rectangular lattice.

For site BB, the six nearest neighbor sites $(13, 14, 22, 24, 33, 34)$ are positioned as follows:

$$
\begin{array}{cccccc}
11 & 12 & \mathbf{\underline{13}} & \mathbf{\underline{14}} & 15 & 16 \\
21 & \mathbf{\underline{22}} & \mathbf{BB} & \mathbf{\underline{24}} & 25 & 26 \\
31 & \mathrm{AA} & \mathbf{\underline{33}} & \mathbf{\underline{34}} & 35 & 36 \\
41 & 42 & 43 & 44 & 45 & 46
\end{array}
$$

In general, the shape of the nearest neighbor sites for a site in an even-numbered row looks like

```
        *   *
    *       *
        *   *
```

Hence, the six nearest neighbors of a site in an even-numbered row on a hexagonal lattice become the nearest neighbors of the site in the N, NE, E, SE, S, W directions on a rectangular lattice.

Using neighborhoods based on the sites identified above for odd- and even-numbered rows, we can develop the algorithm for snowflake formation using a rectangular lattice.

The Snowflake CA

The crystallization lattice begins as a Boolean lattice with m rows and n columns, where m and n are even numbers (while not specifically necessary for the snowflake CA, using even values of m and n makes it possible to employ periodic boundary conditions for other hexagonal lattice CAs). All the sites have value 0 except the center site, which has value 1 (this site serves as the *seed* for the crystallization process)

```
seed = ReplacePart[Table[0, {m}, {n}], 1,{Ceiling[m/2], Ceiling[n/2]}]
```

The rules for snowflake evolution are

- The value of a sites changes to 1 if it has exactly one nearest neighbor site whose value is 1.
- The values of other sites remain unchanged.

These rules can be expressed in the following set of 13 rules:

```
snowflake[a_?OddQ, 0, 1, 0, 0, 0, 0, 0, _, _] := 1
snowflake[a_?OddQ, 0, 0, 1, 0, 0, 0 ,0, _, _] := 1
snowflake[a_?OddQ, 0, 0, 0, 1, 0, 0 ,0, _, _] := 1
```

```
snowflake[a_?OddQ, 0, 0, 0, 0, 1, 0 ,0, _, _] := 1
snowflake[a_?OddQ, 0, 0, 0, 0, 0, 1 ,0, _, _] := 1
snowflake[a_?OddQ, 0, 0, 0, 0, 0, 0 ,1, _, _] := 1

snowflake[a_?EvenQ, 0, 1, 0, 0, 0, _, _, 0, 0] := 1
snowflake[a_?EvenQ, 0, 0, 1, 0, 0, _, _, 0, 0] := 1
snowflake[a_?EvenQ, 0, 0, 0, 1, 0, _, _, 0, 0] := 1
snowflake[a_?EvenQ, 0, 0, 0, 0, 1, _, _, 0, 0] := 1
snowflake[a_?EvenQ, 0, 0, 0, 0, 0, _, _, 1, 0] := 1
snowflake[a_?EvenQ, 0, 0, 0, 0, 0, _, _, 0, 1] := 1

snowflake[_, b_, _, _, _, _, _, _, _, _] := b

Attributes[snowflake] = Listable
```

The 10 arguments to snowflake are, in order,
argument 1: The value of a site in the reference lattice, OddEven,

```
OddEven = Table[i, {i, m}, {n}]
```

which indicates the number of the row that a site in the crystallization lattice lies in (this is used to determine if a site is in an odd- or even-numbered row).

argument 2: The value of a site in the crystallization lattice.

arguments 3–10: The values of the eight nearest neighbor sites in the Moore neighborhood of a site in the crystallization lattice. The matrices of these nearest neighbors can be written as follows, where lat represents the crystallization lattice.

```
RotateRight[lat, {-1, 0}]     - the matrix of nearest neighbor sites in the
                                West direction
RotateRight[lat, {1, 0}]      - the matrix of nearest neighbor sites in the
                                East direction
RotateRight[lat, {0, 1}]      - the matrix of nearest neighbor sites in the
                                North direction
RotateRight[lat, {0, -1}]     - the matrix of nearest neighbor sites in the
                                South direction
RotateRight[lat, {1, 1}]      - the matrix of nearest neighbor sites in the
                                Northeast direction
RotateRight[lat, {-1, 1}]     - the matrix of nearest neighbor sites in the
                                Northwest direction
RotateRight[lat, {1, -1}]     - the matrix of nearest neighbor sites in the
                                Southeast direction
RotateRight[lat, {-1, -1}]    - the matrix of nearest neighbor sites in the
                                Southwest direction
```

Note: For a given site, the values of only two of the last four arguments of snowflake are relevant for updating the site; if the site is in an odd row, the 7th and 8th arguments of snowflake are the relevant sites, and if the site is in an even row, the 9th and 10th arguments of snowflake are the relevant sites.

We can now write the program for snowflake formation.

The Program

```
Snow[m_?EvenQ, n_?EvenQ, t_Integer] :=
Module[{seed, OddEven, snowflake, crystallize},

    seed = ReplacePart[Table[0, {m}, {n}], 1,
                        {Ceiling[m/2], Ceiling[n/2]}];

    snowflake[a_?OddQ, 0, 1, 0, 0, 0, 0, 0, _, _] := 1;
    snowflake[a_?OddQ, 0, 0, 1, 0, 0, 0 ,0, _, _] := 1;
    snowflake[a_?OddQ, 0, 0, 0, 1, 0, 0 ,0, _, _] := 1;
    snowflake[a_?OddQ, 0, 0, 0, 0, 1, 0 ,0, _, _] := 1;
    snowflake[a_?OddQ, 0, 0, 0, 0, 0, 1 ,0, _, _] := 1;
    snowflake[a_?OddQ, 0, 0, 0, 0, 0, 0 ,1, _, _] := 1;

    snowflake[a_?EvenQ, 0, 1, 0, 0, 0, _, _, 0, 0] := 1;
    snowflake[a_?EvenQ, 0, 0, 1, 0, 0, _, _, 0, 0] := 1;
    snowflake[a_?EvenQ, 0, 0, 0, 1, 0, _, _, 0, 0] := 1;
    snowflake[a_?EvenQ, 0, 0, 0, 0, 1, _, _, 0, 0] := 1;
    snowflake[a_?EvenQ, 0, 0, 0, 0, 0, _, _, 1, 0] := 1;
    snowflake[a_?EvenQ, 0, 0, 0, 0, 0, _, _, 0, 1] := 1;

    snowflake[_, b_, _, _, _, _, _, _, _, _] := b;

    Attributes[snowflake] = Listable;

    OddEven = Table[i,{i, m}, {n}];

    crystallize =
      snowflake[OddEven, #,
                RotateRight[#, {-1, 0}], RotateRight[#, {1, 0}],
                RotateRight[#, {0, 1}],  RotateRight[#, {0, -1}],
                RotateRight[#, {1, 1}],  RotateRight[#, {-1, 1}],
                RotateRight[#, {1, -1}], RotateRight[#, {-1, -1}]]&;

    NestList[crystallize, seed, t]
    ]
```

Creating a Hexagonal Lattice Graphic

We can show the snowflake structure after each step of the crystallization process using the graphics program below, SquareToHex, which maps from the rectangular lattice to the hexagonal lattice.

```
SquareToHex[test_]  :=
 Module[{},
  Show[Graphics[
    Table[{Hue[#, #, 1]&[test[[j]][[i]]/Max[Flatten[test]]],
           Polygon[Map[(# + {i 2 Sin[Pi/3] - Mod[j, 2] Sin[Pi/3],
                            3(j - 1) Cos[Pi/3]})&,
                      Table[{Sin[n Pi/3], Cos[n Pi/3]},{n, 0, 5}]]]]},
          {i, Length[test]}, {j, Length[test]}]],
      Frame -> False, AspectRatio -> Automatic,
      DisplayFunction -> Identity]
  ]
```

Here, the evolution of the crystal structure over five time steps is shown.

```
Show[GraphicsArray[Partition[Map[Show[SquareToHex[#]]&,
                    Snow[12, 12, 5]], 2]]];
```

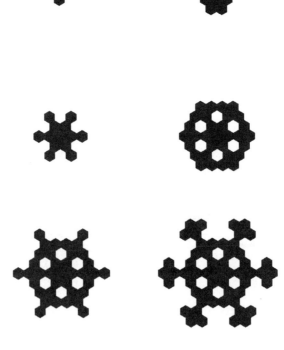

We can superimpose the six graphics above to create a single picture in which the sites added to the crystal at the various time steps are distinguished by their coloring. This is done by adding together the lattices calculated in the Snow program to produce a single list of lattice site values in which the sites that crystallize in the sth time step have the value $(t + 1 - s)$ and these values are taken as arguments to the Hue function in the SquareToHex graphics program.

```
Show[SquareToHex[Apply[Plus, Snow[12, 12, 5]]],
    DisplayFunction -> $DisplayFunction];
```

Computer Simulation Project

Create graphics of snowflake growth over 35 time steps, showing that as the crystal grows its structure cycles through the sequence, plate $\rightarrow$ dendritic arms $\rightarrow$ side branches $\rightarrow$ plate $\rightarrow$ ....

References

Packard, Norman H. "Lattice models for solidification and aggregation." Proceedings of the First International Symposium for Science on Form, Tsukuba University (1986). (Reprinted in Wolfram, Stephen. *Theory and Applications of Cellular Automata*. World Scientific (1986) pp. 305–310.)

7 | Interacting Random Walkers

Introduction

Many apparently disparate processes in nature can be modeled as random walks in which one or more objects move about in space, taking successive steps in randomly chosen directions (see the CAs in the chapters that follow). While a great deal of analytical (equational) work has been done on the properties of a lone random walker, the multiple random walker system is less well analyzed and exact analytical treatments soon become intractable when interactions between the walkers, such as the excluded volume effect, are introduced. We develop here a CA that can be used to model a system of multiple random walkers with excluded volume.

A Single Random Walker

The Single Walker CA

The System

The system consists of a $(2n + 1)$ by $(2n + 1)$ square lattice with periodic boundary conditions. The lattice sites all have value 0 except for one site, initially located at the center of the lattice, whose value is an integer between 1 and 4.

```
initConf  =
ReplacePart[Table[0, {2n + 1}, {2n + 1}], Random[Integer, {1, 4}],
        {n + 1, n + 1}]
```

The site values represent the following quantities:

0 — an empty site
1 — a site occupied by a north-facing walker
2 — a site occupied by an east-facing walker
3 — a site occupied by a south-facing walker
4 — a site occupied by a west-facing walker

The walker always moves in the direction it is facing in a given time step, and hence, a value of 1, 2, 3, 4 indicates that the walker will move in the north, east, south, and west directions, respectively.

The Update Rules

In a given time step, the walker moves from the site it is occupying to one of the four adjacent sites. Accordingly, the rules for updating the lattice sites in the lone walker CA take five arguments. The arguments are, in order, the value of a site, the value of the nearest neighbor site above the site, the value of the nearest neighbor site to the right of the site, the value of the nearest neighbor site below the site, and the value of the nearest neighbor site to the left of the site. The update rules can be stated both in terms of what happens to the walker and in terms of what happens to the sites of the lattice:

- The walker moves from the site it is occupying. The value of an occupied site (i.e., a site with value 1, 2, 3, or 4) beomes 0.

```
walk[1, 0, 0, 0, 0] := 0
walk[2, 0, 0, 0, 0] := 0
walk[3, 0, 0, 0, 0] := 0
walk[4, 0, 0, 0, 0] := 0
```

- The walker moves to the empty adjacent site it is facing and randomly chooses a direction to face. The value of an empty site (i.e., a site with value 0) that is above, right, beneath, or left of a site with value 1, 2, 3, or 4, respectively, is given a randomly selected integer value 1, 2, 3, or 4.

```
walk[0, 3, 0, 0, 0] := RND
walk[0, 0, 4, 0, 0] := RND
walk[0, 0, 0, 1, 0] := RND
walk[0, 0, 0, 0, 2] := RND
```

where

```
RND := Random[Integer, {1, 4}]
```

- An empty site that is not faced by a walker on the adjacent site, remains empty.

```
walk[0, 1, 0, 0, 0] := 0
walk[0, 2, 0, 0, 0] := 0
walk[0, 4, 0, 0, 0] := 0
walk[0, 0, 1, 0, 0] := 0
walk[0, 0, 2, 0, 0] := 0
walk[0, 0, 3, 0, 0] := 0
walk[0, 0, 0, 2, 0] := 0
walk[0, 0, 0, 3, 0] := 0
walk[0, 0, 0, 4, 0] := 0
walk[0, 0, 0, 0, 1] := 0
walk[0, 0, 0, 0, 3] := 0
walk[0, 0, 0, 0, 4] := 0
walk[0, 0, 0, 0, 0] := 0
```

Note: The first four rules above could be combined into a single rule, `walk[_?Positive, 0, 0, 0, 0] := 0` and the last thirteen rules could be combined in the rule, `walk[0, _, _, _, _] := 0`. However, using these compact forms would result in a substantially slower-running program.

Applying the CA Rules

The update rules are applied to the CA lattice using the anonymous function

```
VonNeumann[walk, #]&
```

where

```
VonNeumann[func__, lat_] :=
    MapThread[func, Map[RotateRight[lat, #]&,
              {{0, 0}, {1, 0}, {0, -1}, {-1, 0}, {0, 1}}], 2]
```

The evolution over t time steps is determined using the `NestList` operation.

```
NestList[VonNeumann[walk, #]&,  initConf , t]
```

The Program

```
Meander[n_, t_] :=
 Module[{RND, walk, VonNeumann,    initConf},

  RND := Random[Integer, {1, 4}];

  initConf = ReplacePart[Table[0, {2n + 1}, {2n + 1}], RND,
                         {n + 1, n + 1}];

  walk[1, 0, 0, 0, 0] := 0;
  walk[2, 0, 0, 0, 0] := 0;
  walk[3, 0, 0, 0, 0] := 0;
  walk[4, 0, 0, 0, 0] := 0;

  walk[0, 3, 0, 0, 0] := RND;
  walk[0, 0, 4, 0, 0] := RND;
  walk[0, 0, 0, 1, 0] := RND;
  walk[0, 0, 0, 0, 2] := RND;

  walk[0, 1, 0, 0, 0] := 0;
  walk[0, 2, 0, 0, 0] := 0;
  walk[0, 4, 0, 0, 0] := 0;
  walk[0, 0, 1, 0, 0] := 0;
  walk[0, 0, 2, 0, 0] := 0;
  walk[0, 0, 3, 0, 0] := 0;
  walk[0, 0, 0, 2, 0] := 0;
  walk[0, 0, 0, 3, 0] := 0;
  walk[0, 0, 0, 4, 0] := 0;
  walk[0, 0, 0, 0, 1] := 0;
  walk[0, 0, 0, 0, 3] := 0;
  walk[0, 0, 0, 0, 4] := 0;

  walk[0, 0, 0, 0, 0] := 0;

  VonNeumann[func__, lat_] :=
    MapThread[func, Map[RotateRight[lat, #]&,
              {{0, 0}, {1, 0}, {0, -1}, {-1, 0}, {0, 1}}], 2];

  NestList[VonNeumann[walk, #]&, initConf, t]
 ]
```

Multiple Random Walkers

The Excluded Volume Constraint

In a system of real objects moving about in space, the most basic kind of interaction between the objects is that of volume exclusion, which prohibits having more than one object in a given place at a given time.

Note: Not all multiple-walkers CAs need satisfy the constraints of volume exclusion and/or walker conservation. In some systems (e.g., models of chemical reaction) one or both of these constraints may be inappropriate.

We can create a multiple-walker CA in which the excluded volume constraint is followed (see computer simulation project #1 at the end of the chapter) from the Meander CA simply by placing more than one walker on the lattice initially and then allowing each walker to move to the adjacent site it is facing. However, this will result in a dwindling population of walkers over time because when two or more walkers move to the same site, only one will survive. If we want to conserve the number of walkers in the system over time while imposing the excluded volume constraint, we need to prevent more than one walker from moving to a given lattice site during a time step.

We will show two strategies for conserving the number of walkers while satisfying the volume exclusion constraint.

The Shy Walkers CA

The System

Initially, the sites of an n-by-n square lattice with periodic boundary conditions are randomly occupied by random walkers with probability p, and the values of the occupied sites are randomly selected, using

```
initConf = Table[Floor[p + Random[]], {n}, {n}] *
           Table[Random[Integer, {1, 4}], {n}, {n}]
```

The site values represent the following quantities:

0 — an empty site
1 — a site occupied by a north-facing walker
2 — a site occupied by an east-facing walker
3 — a site occupied by a south-facing walker
4 — a site occupied by a west-facing walker

Note: The density of walkers in `initConf` (which approaches p as the system size increases) is conserved in the system over time, and is given by

```
walkerDensity = N[Count[Flatten[initConf], _?Positive]/n^2]
```

The Update Rules

The volume exclusion constraint is met by having the walkers avoid any confrontation with one another using the following procedure:

In a given time step, each walker in the system moves to the nearest neighbor site lying in the direction the walker is facing if that site is empty, with one exception: when two or more walkers face the same empty site, they remain where they are (e.g., when a right-facing walker lies to the left of an empty site and a south-facing walker lies above the same empty site, both walkers stay put).

The movement of a particle under this avoidance strategy depends on the values of 12 neighboring sites (these sites were shown in the Toolkit chapter) and therefore, the update rules take 13 arguments.

```
walk[site, N, E, S, W, NE, SE, SW, NW, Nn, Ee, Ss, Ww]
```

where the 13 arguments represent the value of the site, the values of the four nearest neighbors in the N, E, S, W directions, the values of the four nearest neighbors in the NE, SE, SW, NW directions, and the values of four next nearest neighbors in the N, E, S, W directions.

The following 28 update rules are used:

At each time step, a walker, regardless of whether it moves or remains in place, randomly chooses a direction to face, using

```
RND := Random[Integer, {1, 4}]
```

- A walker facing an empty site moves from the site it is occupying unless another walker faces the same empty site, in which case it remains in place and randomly chooses a direction to face.

```
walk[1, 0, _, _, _, 4, _, _, _, _, _, _, _] := RND
walk[1, 0, _, _, _, _, _, _, 2, _, _, _, _] := RND
walk[1, 0, _, _, _, _, _, _, 3, _, _, _, _] := RND
walk[1, 0, _, _, _, _, _, _, _, _, _, _, _] := 0
```

```
walk[2, _, 0, _, _, 3, _, _, _, _, _, _, _] := RND
walk[2, _, 0, _, _, _, 1, _, _, _, _, _, _] := RND
walk[2, _, 0, _, _, _, _, _, _, 4, _, _] := RND
walk[2, _, 0, _, _, _, _, _, _, _, _, _] := 0

walk[3, _, _, 0, _, _, 4, _, _, _, _, _, _] := RND
walk[3, _, _, 0, _, _, _, 2, _, _, _, _, _] := RND
walk[3, _, _, 0, _, _, _, _, _, _, 1, _] := RND
walk[3, _, _, 0, _, _, _, _, _, _, _, _] := 0

walk[4, _, _, _, 0, _, _, 1, _, _, _, _, _] := RND
walk[4, _, _, _, 0, _, _, _, 3, _, _, _, _] := RND
walk[4, _, _, _, 0, _, _, _, _, _, _, _, 2] := RND
walk[4, _, _, _, 0, _, _, _, _, _, _, _, _] := 0
```

- Any other walker remains in place and randomly selects a direction
  to face.

```
walk[_?Positive, _, _, _, _,  _, _, _, _, _, _, _, _] := RND
```

- An empty site remains empty if two or more walkers face it.

```
walk[0, 3, 4, _, _, _, _, _, _, _, _, _, _] := 0
walk[0, 3, _, 1, _, _, _, _, _, _, _, _, _] := 0
walk[0, 3, _, _, 2, _, _, _, _, _, _, _, _] := 0
walk[0, _, 4, 1, _, _, _, _, _, _, _, _, _] := 0
walk[0, _, 4, _, 2, _, _, _, _, _, _, _, _] := 0
walk[0, _, _, 1, 2, _, _, _, _, _, _, _, _] := 0
```

- An empty site becomes occupied if it is faced by exactly one walker.

```
walk[0, 3, _, _, _, _, _, _, _, _, _, _, _] := RND
walk[0, _, 4, _, _, _, _, _, _, _, _, _, _] := RND
walk[0, _, _, 1, _, _, _, _, _, _, _, _, _] := RND
walk[0, _, _, _, 2, _, _, _, _, _, _, _, _] := RND
```

- Any other empty site remains empty.

```
walk[0, _, _, _, _, _, _, _, _, _, _, _, _] := 0
```

Applying the Rules

The sites in the lattice are updated at each time step by applying the
following anonymous function to the CA lattice.

```
MvonN[walk, #]&
```

where

```
MvonN[func__, lat_] :=
  MapThread[func, Map[RotateRight[lat, #]&,
            {{0, 0}, {1, 0}, {0, -1}, {-1, 0}, {0, 1},
             {1, -1}, {-1, -1}, {-1, 1}, {1, 1},
             {2, 0}, {0, -2}, {-2, 0}, {0, 2}}], 2]
```

The positions of the random walkers evolve over *t* time steps, starting with the initial lattice configuration, using the following nest operation.

```
NestList[MvonN[walk, #]&, initConf, t]}
```

The Program

```
Hermits[n_, p_, t_] :=
  Module[{walk, initConf, RND, MvonN},

    initConf = Table[Floor[p + Random[]], {n}, {n}] *
               Table[Random[Integer, {1, 4}], {n}, {n}];

    RND := Random[Integer, {1, 4}];

    walk[1, 0, _, _, _, 4, _, _, _, _, _, _, _] := RND;
    walk[1, 0, _, _, _, _, _, _, 2, _, _, _, _] := RND;
    walk[1, 0, _, _, _, _, _, _, 3, _, _, _] := RND;
    walk[1, 0, _, _, _, _, _, _, _, _, _, _] := 0;
    walk[2, _, 0, _, _, 3, _, _, _, _, _, _, _] := RND;
    walk[2, _, 0, _, _, _, 1, _, _, _, _, _, _] := RND;
    walk[2, _, 0, _, _, _, _, _, _, 4, _, _] := RND;
    walk[2, _, 0, _, _, _, _, _, _, _, _, _] := 0;
    walk[3, _, _, 0, _, _, 4, _, _, _, _, _, _] := RND;
    walk[3, _, _, 0, _, _, _, 2, _, _, _, _, _] := RND;
    walk[3, _, _, 0, _, _, _, _, _, _, _, 1, _] := RND;
    walk[3, _, _, 0, _, _, _, _, _, _, _, _] := 0;
    walk[4, _, _, _, 0, _, _, 1, _, _, _, _, _] := RND;
    walk[4, _, _, _, 0, _, _, _, 3, _, _, _, _] := RND;
    walk[4, _, _, _, 0, _, _, _, _, _, _, 2] := RND;
    walk[4, _, _, _, 0, _, _, _, _, _, _, _] := 0;
    walk[_?Positive, _, _, _, _, _, _, _, _, _, _, _, _] := RND;
    walk[0, 3, 4, _, _, _, _, _, _, _, _, _, _] := 0;
    walk[0, 3, _, 1, _, _, _, _, _, _, _, _, _] := 0;
```

```
       walk[0,  3,  _,  _,  2,  _,  _,  _,  _,   _,  _,  _,  _] := 0;
       walk[0,  _,  4,  1,  _,  _,  _,  _,  _,   _,  _,  _,  _] := 0;
       walk[0,  _,  4,  _,  2,  _,  _,  _,  _,   _,  _,  _,  _] := 0;
       walk[0,  _,  _,  1,  2,  _,  _,  _,  _,   _,  _,  _,  _] := 0;
       walk[0,  3,  _,  _,  _,  _,  _,  _,  _,   _,  _,  _,  _] := RND;
       walk[0,  _,  4,  _,  _,  _,  _,  _,  _,   _,  _,  _,  _] := RND;
       walk[0,  _,  _,  1,  _,  _,  _,  _,  _,   _,  _,  _,  _] := RND;
       walk[0,  _,  _,  _,  2,  _,  _,  _,  _,   _,  _,  _,  _] := RND;
       walk[0,  _,  _,  _,  _,  _,  _,  _,  _,   _,  _,  _,  _] := 0;


     MvonN[func__, lat_] :=
       MapThread[func, Map[RotateRight[lat, #]&,
                   {{0, 0}, {1, 0}, {0, -1}, {-1, 0}, {0, 1},
                    {1, -1}, {-1, -1}, {-1, 1}, {1, 1},
                    {2, 0}, {0, -2}, {-2, 0}, {0, 2}}], 2];


       NestList[MvonN[walk, #]& ,  initConf, t]
    ]
```

The Courteous Walkers CA

The System

The system consists of an n-by-n square lattice with periodic boundary conditions whose sites have values that are ordered pairs $\{x, y\}$. The first component of the ordered pair is an integer ranging from 1 to 4 where the values have the following meanings:

0 — an empty site
1 — a site occupied by a north-facing walker
2 — a site occupied by an east-facing walker
3 — a site occupied by a south-facing walker
4 — a site occupied by a west-facing walker

The second component of the ordered pair is a random real number between 0 and 1 where a zero value is associated with an empty site and a nonzero value is associated with a site occupied by a walker. This random number determines if the walker will move or stay put in a time step when it faces an empty site that is also faced by another walker.

We start by randomly placing walkers on the sites of an n-by-n lattice, with a probability p.

```
   initConf = Table[{{Random[Integer, {1, 4}], Random[]} *
                  Floor[Random[] + p], {n}, {n}]
```

Note: The density of walkers in `initConf` (which approaches p as the system size increases) is conserved in the system over time, and is given by

```
walkerDensity = N[Count[Flatten[initConf, 1], {_?Positive, _}]/n^2]
```

The Update Rules

The volume exclusion constraint was dealt with in the Hermits CA by freezing in place any walkers facing the same vacant lattice site on a given time step. An alternative way to deal with this constraint is to use the following procedure:

When two or more walkers face the same empty site, one of the random walkers is randomly selected to move into the empty site while the other walkers remain in place.

This behavior is implemented using the random numbers associated with the walkers.

The rules for updating the lattice sites during a time step take 13 arguments in the following order:

```
walk[site, N, E, S, W, NE, SE, SW, NW, Nn, Ee, Ss, Ww]
```

where the 13 arguments represent the value of the site, the values of the four nearest neighbors in the N, E, S, W directions, the values of the four nearest neighbors in the NE, SE, SW, NW directions, and the values of four next nearest neighbors in the N, E, S, W directions.

At each time step, both the direction faced by a walker and the random number associated with the walker change randomly, regardless of whether the walker moves or remains in place. This is done, using

```
rnd:= {Random[Integer, {1, 4}], Random[]}
```

The behavior of the lattice sites during a time step can be described by the following 38 rules:

• A site occupied by a walker who is facing an empty adjacent site that is also faced by zero or more walkers (eg., a right-facing walker that lies to the left of an empty site that is faced by a left-facing walker lying to the right of the empty site) remains occupied and randomly selects a new associated random number and direction to face, if the walker's associated random number is less than any of the random numbers associated with the other walkers facing the same site. If the

walker's associated random number is greater than all of the other
random numbers, the site becomes empty as the walker moves.

```
walk[{1, a_}, {0, 0}, _, _, _, {4, b_},
        _, _, {2, c_}, {3, d_}, _, _, _] := rnd /; a != Max[a, b, c, d]
walk[{1, a_}, {0, 0}, _, _, _, {4, b_}, _, _, {2, c_}, _, _, _, _] :=
                                        rnd /; a != Max[a, b, c]
walk[{1, a_}, {0, 0}, _, _, _, {4, b_}, _, _, _, {3, d_}, _, _, _] :=
                                        rnd /; a != Max[a, b, d]
walk[{1, a_}, {0, 0}, _, _, _, _, _, _, {2, c_}, {3, d_}, _, _, _] :=
                                        rnd /; a != Max[a, c, d]
walk[{1, a_}, {0, 0}, _, _, _, {4, b_}, _, _, _, _, _, _, _] :=
                                        rnd /; a != Max[a, b]
walk[{1, a_}, {0, 0}, _, _, _, _, _, _, {2, c_}, _, _, _, _] :=
                                        rnd /; a != Max[a, c]
walk[{1, a_}, {0, 0}, _, _, _, _, _, _, _, {3, d_}, _, _, _] :=
                                        rnd /; a != Max[a, d]
walk[{1, a_}, {0, 0}, _, _, _, _, _, _, _, _, _, _, _] := {0, 0}

walk[{2, a_}, _, {0, 0}, _, _, {3, b_},
        {1, c_}, _, _, _, {4, d_}, _, _] := rnd /; a != Max[a, b, c, d]
walk[{2, a_}, _, {0, 0}, _, _, {3, b_}, {1, c_}, _, _, _, _, _, _] :=
                                        rnd /; a != Max[a, b, c]
walk[{2, a_}, _, {0, 0}, _, _, {3, b_}, _, _, _, _, {4, d_}, _, _] :=
                                        rnd /; a != Max[a, b, d]
walk[{2, a_}, _, {0, 0}, _, _, _, {1, c_}, _, _, _, {4, d_}, _, _] :=
                                        rnd /; a != Max[a, c, d]
walk[{2, a_}, _, {0, 0}, _, _, {3, b_}, _, _, _, _, _, _, _] :=
                                        rnd /; a != Max[a, b]
walk[{2, a_}, _, {0, 0}, _, _, _, {1, c_}, _, _, _, _, _, _] :=
                                        rnd /; a != Max[a, c]
walk[{2, a_}, _, {0, 0}, _, _, _, _, _, _, _, {4, d_}, _, _] :=
                                        rnd /; a != Max[a, d]
walk[{2, a_}, _, {0, 0}, _, _, _, _, _, _, _, _, _, _] := {0, 0}

walk[{3, a_}, _, _, {0, 0}, _, _, {4, b_},
        {2, c_}, _, _, _, {1, d_}, _] := rnd /; a != Max[a, b, c, d]
walk[{3, a_}, _, _, {0, 0}, _, _, {4, b_}, {2, c_}, _, _, _, _, _] :=
                                        rnd /; a != Max[a, b, c]
walk[{3, a_}, _, _, {0, 0}, _, _, {4, b_}, _, _, _, _, {1, d_}, _] :=
                                        rnd /; a != Max[a, b, d]
walk[{3, a_}, _, _, {0, 0}, _, _,_, {2, c_}, _, _, _, {1, d_}, _] :=
                                        rnd /; a != Max[a, c, d]
walk[{3, a_}, _, _, {0, 0}, _, _, {4, b_}, _, _, _, _, _, _] :=
                                        rnd /; a != Max[a, b]
```

```
walk[{3, a_}, _, _, {0, 0}, _, _, _, {2, c_}, _, _, _, _, _] :=
                                            rnd /; a != Max[a, c]
walk[{3, a_}, _, _, {0, 0}, _, _, _, _, _, _, _, {1, d_}, _] :=
                                            rnd /; a != Max[a, d]
walk[{3, a_}, _, _, {0, 0}, _, _, _, _, _, _, _, _, _] := {0, 0}


walk[{4, a_}, _, _, _, {0, 0}, _, _, {1, b_},
            {3, c_}, _, _, _, {2, d_}] := rnd /; a != Max[a, b, c, d]
walk[{4, a_}, _, _, _, {0, 0}, _, _, {1, b_}, {3, c_}, _, _, _, _] :=
                                            rnd /; a != Max[a, b, c]
walk[{4, a_}, _, _, _, {0, 0}, _, _, {1, b_}, _, _, _, _, {2, d_}] :=
                                            rnd /; a != Max[a, b, d]
walk[{4, a_}, _, _, _, {0, 0}, _, _, _, {3, c_}, _, _, _, {2, d_}] :=
                                            rnd /; a != Max[a, c, d]
walk[{4, a_}, _, _, _, {0, 0}, _, _, {1, b_}, _, _, _, _, _] :=
                                            rnd /; a != Max[a, b]
walk[{4, a_}, _, _, _, {0, 0}, _, _, _, {3, c_}, _, _, _, _] :=
                                            rnd /; a != Max[a, c]
walk[{4, a_}, _, _, _, {0, 0}, _, _, _, _, _, _, _, {2, d_}] :=
                                            rnd /; a != Max[a, d]
walk[{4, a_}, _, _, _, {0, 0}, _, _, _, _, _, _, _, _] := {0, 0}
```

- A site occupied by a walker who is facing an occupied site remains oc-cupied by the walker who randomly selects a new associated random number and direction to face.

```
walk[{_?Positive, _}, _, _, _, _, _,_, _, _, _, _, _, _] := rnd
```

- An empty site faced by one or more walkers becomes occupied by a walker who randomly selects a new associated random number and direction to face.

```
walk[{0, 0}, {3, _}, _, _, _, _, _, _, _, _, _, _, _] := rnd
walk[{0, 0}, _, {4, _}, _, _, _, _, _, _, _, _, _, _] := rnd
walk[{0, 0}, _, _, {1, _}, _, _, _, _, _, _, _, _, _] := rnd
walk[{0, 0}, _, _, _, {2, _}, _, _, _, _, _, _, _, _] := rnd
```

- An empty site that is not faced by a walker remains empty.

```
walk[{0, 0}, _, _, _, _, _, _, _, _, _, _, _, _] := {0, 0}
```

Applying the Rules

The sites in the lattice are updated at each time step by applying the anonymous MvonN function defined earlier, to the lattice.

The positions of the random walkers evolve over *t* time steps, starting with the initial lattice configuration, using the following nest operation.

```
NestList[MvonN[walk, #]& ,  initConf, t]
```

The Program

```
Gentlemen[n_, p_, t_]:=
Module[{initConf, walk, rnd, MvonN},

 rnd := {Random[Integer, {1, 4}], Random[]};

 initConf = Table[rnd* Floor[Random[] + p], {n}, {n}];

 walk[{1, a_}, {0, 0}, _, _, _, {4, b_},
       _, _, {2, c_}, {3, d_}, _, _, _] := rnd /; a != Max[a, b, c, d];
 walk[{1, a_}, {0, 0}, _, _, _, {4, b_}, _, _, {2, c_}, _, _, _, _] :=
                                   rnd /; a != Max[a, b, c];
 walk[{1, a_}, {0, 0}, _, _, _, {4, b_}, _, _, _, {3, d_}, _, _, _] :=
                                   rnd /; a != Max[a, b, d];
 walk[{1, a_}, {0, 0}, _, _, _, _, _, _, {2, c_}, {3, d_}, _, _, _] :=
                                   rnd /; a != Max[a, c, d];
 walk[{1, a_}, {0, 0}, _, _, _, {4, b_}, _, _, _, _, _, _, _] :=
                                   rnd /; a != Max[a, b];
 walk[{1, a_}, {0, 0}, _, _, _, _, _, _, {2, c_}, _, _, _, _] :=
                                   rnd /; a != Max[a, c];
 walk[{1, a_}, {0, 0}, _, _, _, _, _, _, _, {3, d_}, _, _, _] :=
                                   rnd /; a != Max[a, d];
 walk[{1, a_}, {0, 0}, _, _, _, _, _, _, _, _, _, _, _] := {0, 0};
 walk[{2, a_}, _, {0, 0}, _, _, {3, b_},
       {1, c_}, _, _, _, {4, d_}, _, _] := rnd /; a != Max[a, b, c, d];
 walk[{2, a_}, _, {0, 0}, _, _, {3, b_}, {1, c_}, _, _, _, _, _, _] :=
                                   rnd /; a != Max[a, b, c];
 walk[{2, a_}, _, {0, 0}, _, _, {3, b_}, _, _, _, _, {4, d_}, _, _] :=
                                   rnd /; a != Max[a, b, d];
 walk[{2, a_}, _, {0, 0}, _, _, _, {1, c_}, _, _, _, {4, d_}, _, _] :=
                                   rnd /; a != Max[a, c, d];
 walk[{2, a_}, _, {0, 0}, _, _, {3, b_}, _, _, _, _, _, _, _] :=
                                   rnd /; a != Max[a, b];
 walk[{2, a_}, _, {0, 0}, _, _, _, {1, c_}, _, _, _, _, _, _] :=
                                   rnd /; a != Max[a, c];
 walk[{2, a_}, _, {0, 0}, _, _, _, _, _, _, _, {4, d_}, _, _] :=
                                   rnd /; a != Max[a, d];
 walk[{2, a_}, _, {0, 0}, _, _, _, _, _, _, _, _, _, _] := {0, 0};
```

```
walk[{3, a_}, _, _, {0, 0}, _, _, {4, b_},
        {2, c_}, _, _, _, {1, d_}, _] := rnd /; a != Max[a, b, c, d];
walk[{3, a_}, _, _, {0, 0}, _, _, {4, b_}, {2, c_}, _, _, _, _, _] :=
                                        rnd /; a != Max[a, b, c];
walk[{3, a_}, _, _, {0, 0}, _, _, {4, b_}, _, _, _, _, {1, d_}, _] :=
                                        rnd /; a != Max[a, b, d];
walk[{3, a_}, _, _, {0, 0}, _, _,_, {2, c_}, _, _, _, {1, d_}, _] :=
                                        rnd /; a != Max[a, c, d];
walk[{3, a_}, _, _, {0, 0}, _, _, {4, b_}, _, _, _, _, _, _] :=
                                        rnd /; a != Max[a, b];
walk[{3, a_}, _, _, {0, 0}, _, _, _, {2, c_}, _, _, _, _, _] :=
                                        rnd /; a != Max[a, c];
walk[{3, a_}, _, _, {0, 0}, _, _, _, _, _, _, _, {1, d_}, _] :=
                                        rnd /; a != Max[a, d];
walk[{3, a_}, _, _, {0, 0}, _, _, _, _, _, _, _, _, _] := {0, 0};
walk[{4, a_}, _, _, _, {0, 0}, _, _, {1, b_},
        {3, c_}, _, _, _, {2, d_}] := rnd /; a != Max[a, b, c, d];
walk[{4, a_}, _, _, _, {0, 0}, _, _, {1, b_}, {3, c_}, _, _, _, _] :=
                                        rnd /; a != Max[a, b, c];
walk[{4, a_}, _, _, _, {0, 0}, _, _, {1, b_}, _, _, _, _, {2, d_}] :=
                                        rnd /; a != Max[a, b, d];
walk[{4, a_}, _, _, _, {0, 0}, _, _, _, {3, c_}, _, _, _, {2, d_}] :=
                                        rnd /; a != Max[a, c, d];
walk[{4, a_}, _, _, _, {0, 0}, _, _, {1, b_}, _, _, _, _, _] :=
                                        rnd /; a != Max[a, b];
walk[{4, a_}, _, _, _, {0, 0}, _, _, _, {3, c_}, _, _, _, _] :=
                                        rnd /; a != Max[a, c];
walk[{4, a_}, _, _, _, {0, 0}, _, _, _, _, _, _, _, {2, d_}] :=
                                        rnd /; a != Max[a, d];
walk[{4, a_}, _, _, _, {0, 0}, _, _, _, _, _, _, _, _] := {0, 0};
walk[{_?Positive, _}, _, _, _, _, _,_, _, _, _, _, _, _] := rnd;
walk[{0, 0}, {3, _}, _, _, _, _, _, _, _, _, _, _, _] := rnd;
walk[{0, 0}, _, {4, _}, _, _, _, _, _, _, _, _, _, _] := rnd;
walk[{0, 0}, _, _, {1, _}, _, _, _, _, _, _, _, _, _] := rnd;
walk[{0, 0}, _, _, _, {2, _}, _, _, _, _, _, _, _, _] := rnd;
walk[{0, 0}, _, _, _, _, _, _, _, _, _, _, _, _] := {0, 0};

MvonN[func__, lat_] :=
  MapThread[func, Map[RotateRight[lat, #]&,
            {{0, 0}, {1, 0}, {0, -1}, {-1, 0}, {0, 1},
             {1, -1}, {-1, -1}, {-1, 1}, {1, 1},
             {2, 0}, {0, -2}, {-2, 0}, {0, 2}}]], 2];

NestList[MvonN[walk, #]&, initConf, t]
]
```

Computer Simulation Projects

1. Modify the `Meander` program for the single walker by placing more than one walker on the lattice initially and allowing each walker in the many-walker system to move to the adjacent site it is facing regardless of whether that site is occupied or empty. Run the program and determine the density of walkers on the lattice over time.

2. An interesting system of interacting random walkers, known as the *vicious* walkers model, consists of walkers who "shoot each other dead on sight" so that when more than one walker reaches the same site at the same time, they are all eliminated (this behavior is reminiscent of the final scene in the movie *Reservoir Dogs*). Modify the `Hermits` program to model a system of *N* vicious walkers.

3. The expansion of a system of particles (e.g., gas molecules) with volume exclusion can be modeled using the Shy Random Walkers CA. The only modifications that need to be made to the `Hermits` program are (1) decorate the initial lattice configuration with a border of empty sites, and (2) apply the following anonymous conditional function to the lattice that results from the application of the `walk` function to the lattice in each time step.

```
increaseLat =
 If[Max[Map[{Max[First[#]], Max[Last[#]]}&,
            {#, Transpose[#]}&[#]]] != 0,
    zeros[#], #]&
```

where

```
zeros = (Prepend[Append[Map[Prepend[Append[#, 0], 0]&, #],
                       Table[0, {Length[#] + 2}]],
                Table[0, {Length[#] + 2}]])&
```

The *zeros* function adds a border of zeros to the lattice (see the Solidification CA). *increaseLat* is used because the periodic boundary conditions results in particles disappearing off one side of the lattice and reappearing on the opposite side of the lattice and adding the borders of 0's whenever a particle moves to a site along the border, prevents this.

Write a CA program for the expansion of a system of particles.

4. The lattices created by the previous project increase over time because a border of 0's is added to the lattice at various time steps. To make an animation of the expansion process we need to take the list of lattice configurations generated by the program and add enough rows and

columns of 0's to each element in the list to make all of the lattices the same size. This can be done using the following anonymous function:

```
lis = Function[y, Map[Nest[zeros, y[[#]], Length[y] - #]&,
                  Range[Length[y]]]]
```

Create an animation using the previous program.

5. The mean number of distinct sites visited by a random walker (the number of sites visited at least once) as a function of time is relevant to various phenomena in materials, chemistry, physics, and ecology (Shlesinger, 1992). It is also relevant to a system of many random walkers (e.g., the territory explored by members of an animal species that start off near each other is given by this quantity). The basic computation involves determining the mean number of sites that are visited for the first time by the random walkers in each time step (the occupation of a site for the first time is known as a first passage event).

An analytical calculation of the number of distinct sites visited as a function of time has been made (Larraide, et al., 1992) for the simple case of N "harmless" random walkers (this is equivalent to an "ideal gas" system of noninteracting walkers who can occupy the same site at the same time). The results indicate that for a large number of walkers who are all initially localized at the origin, the power law behavior of the mean number of distinct sites visited changes over three time regimes [varying at very short times as t^2, at intermediate times as $t * \ln(N/\ln(t))$, and at long times as $t * N/\ln(t)$].

The list of lattice configurations, lis, obtained from running the expansion program given in the first computer simulation project, can be used to calculate the mean number of distinct sites visited as a function of time for the case of N shy random walkers, each of whom avoids confrontations by not moving to the same place as another walker at the same time. The computation is done by applying the following function to expansion[n, p, t]:

```
numberDistinctSites2D[x_] :=
  Map[Length, Rest[FoldList[Union, {}, Map[Position[#, _?Positive]&,
      x]]]]
```

A plot of the number of the distinct sites visited by N "shy" walkers vs. time can then be drawn using

```
ListPlot[numberDistinctSites2D[lis], PlotStyle -> PointSize[.03]]
```

Calculate the power law behavior of a system of N "shy" walkers and compare it with the behavior found for the system of N *harmless* walkers. To help visualize your results, create graphics of the set of distinct sites visited by N shy walkers at short, intermediate, and long times and notice how the shape and roughness of the surface of the set varies over time.

References

Larraide, Hernana, Trunfio, Paul, Havlin, Shlomo, Stanley, Eugene H., and Weiss, George H. 1992. "Territory covered by N diffusing particles." Nature, 355, 423–426.
Shlesinger, Michael F. 1992. "New paths for random walkers." Nature, 355, 396–397.

8 Interfacial Diffusion Fronts and Gradient Percolation

Introduction

When two materials are placed in contact with one another, an interface forms between them. The nature of the interface is important in determining a variety of properties of the system, such as adhesion strength, mechanical strength, thermal expansion, electrical conductivity, etc.

The interdiffusion of the atoms of the two materials results in the formation of clusters of A atoms and clusters of B atoms interfacing each other. Some of the clusters of atoms of a given type will be isolated, surrounded by atoms of the other type, but there will be one *infinite* cluster for each type of atom emanating from the diffusion source for atoms of that type. The *diffusion front* consists of those atoms that interface between these two infinite clusters.

We will first present two methods for generating an interfacial diffusion front and then show how to determine the diffusion front that forms.

Generating a Diffusion Front

We'll consider the diffusion of particles A on a two-dimensional lattice B with the diffusion source being a line of A atoms kept a constant concentration of one. The concentration of A particles will vary as a function of time and the distance from the source. The diffusion front will be generated in two ways: dynamically using the random walkers CA, and statically using a gradient percolation approach.

Diffusion CA

The system consists of an $n+1$-by-n lattice. Each lattice site is either empty or is occupied by a particle. The lattice site values are nonnegative integer values that represent the following quantities:

0 — an empty site

1 — a site occupied by a north-facing particle

2 — a site occupied by an east-facing particle

3 — a site occupied by a south-facing particle

4 — a site occupied by a west-facing particle

Initially, all of the sites in the lattice are empty (i.e., the sites have value 0), except for the sites in the bottom row, which are occupied by particles facing randomly chosen directions (i.e., these sites have positive integer values).

The left and right boundaries of the lattice are periodic. The bottom boundary acts like a source, so that a particle moving upward from the bottom row is replaced by another particle. The top boundary acts like a sink, so that a particle moving upward from the top row vanishes. To implement the sink and source, we attach to the bottom of the lattice, a row whose sites have the symbolic value b.

Overall, the lattice is created using

```
initConf = Join[Table[0, {n}, {n}],
                {Table[Random[Integer, {1, 4}], {n}]},
                {Table[b, {n}]}]
```

The system evolves over a succession of time steps. In each time step, each particle moves to the nearest neighbor site lying in the direction the particle is facing (hence, a value of 1, 2, 3, 4 indicates that the particle moves in the north, east, south, and west directions, respectively) if the site is empty and if the move will not result in a collision with another particle moving into the same site; otherwise, it remains in place. Whenever a particle moves out of the next-to-last row, it is replaced by another particle, and whenever a particle moves upward from the first row, it disappears. The bottom row of bs remains unchanged.

The program for the diffusion CA is the same as the program used for the interacting random walker CA with three additional rules (these are the last three update rules in the program that follows, each one preceded by a parenthetical comment explaining its use).

```
interfacialDiffusion[n_, t_] :=
 Module[{walk, initConf, RND, MvonN},

 initConf = Join[Table[0, {n}, {n}],
                 {Table[Random[Integer, {1, 4}], {n}]},
                 {Table[b, {n}]}];

 RND := Random[Integer, {1, 4}];
```

```
walk[1, 0, _, _, _, 4, _, _, _, _, _, _, _] := RND;
walk[1, 0, _, _, _, _, _, _, 2, _, _, _, _] := RND;
walk[1, 0, _, _, _, _, _, _, _, 3, _, _, _] := RND;
walk[1, 0, _, _, _, _, _, _, _, _, _, _, _] := 0;
walk[2, _, 0, _, _, 3, _, _, _, _, _, _, _] := RND;
walk[2, _, 0, _, _, _, 1, _, _, _, _, _, _] := RND;
walk[2, _, 0, _, _, _, _, _, _, 4, _, _] := RND;
walk[2, _, 0, _, _, _, _, _, _, _, _, _] := 0;
walk[3, _, _, 0, _, _, 4, _, _, _, _, _, _] := RND;
walk[3, _, _, 0, _, _, _, 2, _, _, _, _, _] := RND;
walk[3, _, _, 0, _, _, _, _, _, _, 1, _] := RND;
walk[3, _, _, 0, _, _, _, _, _, _, _, _] := 0;
walk[4, _, _, _, 0, _, _, _, 1, _, _, _, _] := RND;
walk[4, _, _, _, 0, _, _, _, 3, _, _, _, _] := RND;
walk[4, _, _, _, 0, _, _, _, _, _, _, _, 2] := RND;
walk[4, _, _, _, 0, _, _, _, _, _, _, _, _] := 0;
walk[_?Positive, _, _, _, _, _, _, _, _, _, _, _, _] := RND;
walk[0, 3, 4, _, _, _, _, _, _, _, _, _, _] := 0;
walk[0, 3, _, 1, _, _, _, _, _, _, _, _, _] := 0;
walk[0, 3, _, _, 2, _, _, _, _, _, _, _, _] := 0;
walk[0, _, 4, 1, _, _, _, _, _, _, _, _, _] := 0;
walk[0, _, 4, _, 2, _, _, _, _, _, _, _, _] := 0;
walk[0, _, _, 1, 2, _, _, _, _, _, _, _, _] := 0;
walk[0, 3, _, _, _, _, _, _, _, _, _, _, _] := RND;
walk[0, _, 4, _, _, _, _, _, _, _, _, _, _] := RND;
walk[0, _, _, 1, _, _, _, _, _, _, _, _, _] := RND;
walk[0, _, _, _, 2, _, _, _, _, _, _, _, _] := RND;
walk[0, _, _, _, _, _, _, _, _, _, _, _, _] := 0;

(* a site in the border row remains unchanged *)
walk[b, _, _, _, _, _, _, _, _, _, _, _, _] := b;

(* a particle moving out of the row above the border row is replaced
   by another particle *)
walk[1, 0, _, b, _, _, _, _, _, _, _, _, _] := RND;

(* a particle moving upward from the top row disappears *)
walk[1, b, _, _, _, _, _, _, _, _, _, _, _] := 0;

MvonN[func__, lat_] :=
  MapThread[func, Map[RotateRight[lat, #]&,
            {{0, 0}, {1, 0}, {0, -1}, {-1, 0}, {0, 1},
             {1, -1}, {-1, -1}, {-1, 1}, {1, 1},
             {2, 0}, {0, -2}, {-2, 0}, {0, 2}}], 2];

NestList[MvonN[walk, #]&, initConf, t]
]
```

Gradient Percolation

While the diffusion CA program above is useful for studying the dynamics of an interfacial diffusion process, for static properties it suffices to simply create a lattice with a gradient of concentration of occupied sites (Sapoval, et al., 1985). This is done by distributing particles randomly on each row of the lattice with a density distribution probability p that decreases monotonically with the distance of a row from the first row. The precise form of p is not important (e.g., the concentration of particles can have a linear or Erfc dependence on position) as long as the directional dependence fills the first row and leaves the last row empty. This is known as a *gradient percolation* system.

Note: The gradient percolation system differs from the traditional random site percolation model, which is a random Boolean lattice generated using

```
sitePercolation[p_, m_] := Table[Floor[1 + p - Random[]], {m}, {m}]
```

where p, the probability of a lattice site being occupied, is constant over the whole lattice (i.e., the same for all sites in the lattice).

The gradient percolation system consists of a rectangular lattice having m rows and n columns. The probability of a site being occupied in a given row decreases linearly with the distance of the row from the first row.

The lattice can be generated using the following program:

```
gradientPercolation[n_, m_] :=
 Module[{p = 0},
   createRow = (p += 1/(n - 1);
               h[#, Table[Floor[Random[] + p], {m}]])&;
   Apply[List,
       Flatten[Nest[createRow, Table[0, {m}], n - 1], Infinity, h]]
     ]
```

A typical gradient percolation system is shown below.

```
Show[Graphics[RasterArray[gradientPercolation[100, 100] /.
                {0 -> RGBColor[0, 1, 0],1 -> RGBColor[0, 0, 1]}]],
      AspectRatio->Automatic];
```

Identifying the Diffusion Front

A diffusion front (or "seashore" or "frontier") is defined (Sapoval, et al., 1985) by looking at the cluster of occupied (or A) sites containing the occupied sites in the last row of the lattice and the cluster of empty (or B) sites containing the empty sites in the first row. The interface consisting of the sites that give the closest possible contact between these two *infinite* clusters can be obtained by determining the sites in these two clusters in different ways:

The sites in the cluster of occupied sites are in the von Neumann neighborhood of other sites in the cluster (the von Neumann neighborhood consists of the four nearest neighbor sites lying N, E, S, W of a site).

The sites in the cluster of empty sites are in the Moore neighborhood of other sites in the cluster (the Moore neighborhood consists of the eight nearest neighbor sites lying N, NE, E, SE, S, SW, W, NW of a site).

Two diffusion fronts can be defined (Chopard, et al., 1989):

(1) The sites in the infinite empty site cluster that lie in the von Neumann neighborhood of a site in the infinite occupied site cluster.

(2) The sites in the infinite occupied site cluster that lie in the Moore neighborhood of a site in the infinite empty site cluster.

In determining the diffusion fronts, the lattice is taken to have periodic boundary conditions in the direction perpendicular to the concentration gradient.

We will show how to determine the diffusion front of a two-dimensional gradient percolation system in a step-by-step fashion. Below is the simple system that we will use to illustrate the computation. The occupied sites have value 1 and the unoccupied sites have value 0.

```
sapoval =
 {{0, 0, 0, 0, 0, 0, 0, 0, 0, 0, 0, 0, 0, 0, 0, 0, 0},
  {0, 1, 1, 0, 0, 0, 0, 0, 0, 0, 0, 0, 0, 0, 0, 0, 0},
  {0, 0, 0, 0, 0, 0, 0, 0, 0, 0, 1, 0, 0, 0, 0, 0, 0},
  {0, 0, 1, 0, 0, 1, 0, 0, 1, 0, 1, 1, 0, 0, 0, 0, 0},
  {0, 0, 1, 0, 0, 1, 0, 0, 1, 0, 0, 0, 0, 1, 0, 0, 0},
  {1, 0, 0, 0, 1, 0, 0, 0, 0, 1, 1, 0, 0, 0, 0, 1, 0},
  {0, 0, 0, 1, 1, 0, 0, 0, 1, 1, 1, 0, 1, 0, 0, 0, 0},
  {0, 1, 0, 0, 1, 0, 0, 1, 0, 0, 0, 0, 1, 0, 0, 0, 0},
  {1, 1, 1, 0, 0, 0, 0, 0, 1, 1, 0, 1, 1, 0, 1, 0, 0},
  {1, 1, 1, 1, 1, 1, 1, 1, 0, 1, 0, 0, 0, 1, 1, 0, 1},
  {1, 1, 1, 1, 0, 0, 0, 0, 1, 1, 0, 1, 0, 1, 1, 1, 1},
  {1, 1, 1, 1, 0, 1, 1, 0, 1, 1, 1, 1, 0, 1, 1, 1, 0},
  {1, 1, 1, 1, 0, 0, 1, 1, 0, 1, 1, 1, 1, 1, 0, 1, 0},
  {1, 1, 1, 1, 1, 1, 1, 1, 1, 1, 0, 1, 1, 1, 1, 1, 1},
  {1, 0, 0, 1, 1, 1, 1, 1, 0, 1, 1, 1, 1, 1, 1, 1, 1},
  {1, 1, 1, 1, 1, 1, 1, 1, 1, 1, 1, 1, 1, 1, 1, 1, 1},
  {1, 1, 1, 1, 1, 1, 1, 1, 1, 1, 1, 1, 1, 1, 1, 1, 1}};
```

We first identify the clusters of occupied sites in the sapoval lattice. This can be done using the cluster labeling program (see the Excitable Media Contagion chapter for the development of this code).

```
clusterLabelvonNeumann[mat_List]:=
 Module[{i = 2, clusterCornerID, cornerLabels, reLabel, VonNeumann},
  clusterCornerID[1,0, 0]:= i++;
  clusterCornerID[a_, __]:= a;
  Attributes[clusterCornerID] = Listable;

  cornerLabels = clusterCornerID[#, RotateRight[#,{1,0}],
                          RotateRight[#,{0, 1}]]&[mat];

  reLabel[0, ___] := 0;
  reLabel[a_, b_, c_, d_, e_] := Max[a, b, c, d, e];
```

```
VonNeumann[func__, lat_] :=
    MapThread[func, Map[RotateRight[lat, #]&,
            {{0, 0}, {1, 0}, {0, -1}, {-1, 0}, {0, 1}}], 2];

FixedPoint[VonNeumann[reLabel, #]&, cornerLabels]
]
```

Applying the `clusterLabelvonNeumann` function to the sapoval lattice and then renumbering the identified clusters so that their numbers run sequentially with no gaps, we get

```
AClustersLat = clusterLabelvonNeumann[sapoval];

AClusters =
 AClustersLat /.
  MapThread[Rule,
            {Reverse[Rest[Union[Flatten[AClustersLat]]]],
             Range[Length[Union[Flatten[AClustersLat]]] - 1]}
        ]
```

```
{{0, 0, 0, 0, 0, 0, 0, 0, 0, 0, 0, 0, 0, 0, 0, 0, 0},
 {0, 13, 13, 0, 0, 0, 0, 0, 0, 0, 0, 0, 0, 0, 0, 0, 0},
 {0, 0, 0, 0, 0, 0, 0, 0, 0, 0, 12, 0, 0, 0, 0, 0, 0},
 {0, 0, 11, 0, 0, 10, 0, 0, 9, 0, 12, 12, 0, 0, 0, 0, 0},
 {0, 0, 11, 0, 0, 10, 0, 0, 9, 0, 0, 0, 0, 8, 0, 0, 0},
 {7, 0, 0, 0, 5, 0, 0, 0, 0, 4, 4, 0, 0, 0, 0, 6, 0},
 {0, 0, 0, 5, 5, 0, 0, 0, 4, 4, 4, 0, 2, 0, 0, 0, 0},
 {0, 1, 0, 0, 5, 0, 0, 3, 0, 0, 0, 0, 2, 0, 0, 0, 0},
 {1, 1, 1, 0, 0, 0, 0, 0, 1, 1, 0, 2, 2, 0, 1, 0, 0},
 {1, 1, 1, 1, 1, 1, 1, 1, 0, 1, 0, 0, 0, 1, 1, 0, 1},
 {1, 1, 1, 1, 0, 0, 0, 0, 1, 1, 0, 1, 0, 1, 1, 1, 1},
 {1, 1, 1, 1, 0, 1, 1, 0, 1, 1, 1, 1, 0, 1, 1, 1, 0},
 {1, 1, 1, 1, 0, 0, 1, 1, 0, 1, 1, 1, 1, 0, 1, 0},
 {1, 1, 1, 1, 1, 1, 1, 1, 1, 1, 0, 1, 1, 1, 1, 1, 1},
 {1, 0, 0, 1, 1, 1, 1, 1, 0, 1, 1, 1, 1, 1, 1, 1, 1},
 {1, 1, 1, 1, 1, 1, 1, 1, 1, 1, 1, 1, 1, 1, 1, 1, 1},
 {1, 1, 1, 1, 1, 1, 1, 1, 1, 1, 1, 1, 1, 1, 1, 1, 1}}
```

The `clusterLabelvonNeumann` program labels clusters of connected occupied sites (i.e., sites that have value 1 and lie in the von Neumann neighborhood of other sites that have value 1), starting in the bottom row and proceeding upward. Therefore, the sites having value 1 in AClusters belong to the infinite occupied site cluster. Their locations are determined using the `Position` function.

```
infiniteAClusterlocs = Position[AClusters, 1]
```

```
{{8, 2}, {9, 1}, {9, 2}, {9, 3}, {9, 9}, {9, 10}, {9, 15},
 {10, 1}, {10, 2}, {10, 3}, {10, 4}, {10, 5}, {10, 6},
 {10, 7}, {10, 8}, {10, 10}, {10, 14}, {10, 15}, {10, 17},
 {11, 1}, {11, 2}, {11, 3}, {11, 4}, {11, 9}, {11, 10},
 {11, 12}, {11, 14}, {11, 15}, {11, 16}, {11, 17}, {12, 1},
 {12, 2}, {12, 3}, {12, 4}, {12, 6}, {12, 7}, {12, 9},
 {12, 10}, {12, 11}, {12, 12}, {12, 14}, {12, 15}, {12, 16},
 {13, 1}, {13, 2}, {13, 3}, {13, 4}, {13, 7}, {13, 8},
 {13, 10}, {13, 11}, {13, 12}, {13, 13}, {13, 14}, {13, 16},
 {14, 1}, {14, 2}, {14, 3}, {14, 4}, {14, 5}, {14, 6},
 {14, 7}, {14, 8}, {14, 9}, {14, 10}, {14, 12}, {14, 13},
 {14, 14}, {14, 15}, {14, 16}, {14, 17}, {15, 1}, {15, 4},
 {15, 5}, {15, 6}, {15, 7}, {15, 8}, {15, 10}, {15, 11},
 {15, 12}, {15, 13}, {15, 14}, {15, 15}, {15, 16}, {15, 17},
 {16, 1}, {16, 2}, {16, 3}, {16, 4}, {16, 5}, {16, 6},
 {16, 7}, {16, 8}, {16, 9}, {16, 10}, {16, 11}, {16, 12},
 {16, 13}, {16, 14}, {16, 15}, {16, 16}, {16, 17}, {17, 1},
 {17, 2}, {17, 3}, {17, 4}, {17, 5}, {17, 6}, {17, 7},
 {17, 8}, {17, 9}, {17, 10}, {17, 11}, {17, 12}, {17, 13},
 {17, 14}, {17, 15}, {17, 16}, {17, 17}}
```

To identify the clusters of unoccupied sites, we modify the cluster-LabelvonNeumann program so that it labels connected sites that have value 1 and lie in the Moore neighborhood of other sites that have value 1.

```
clusterLabelMoore[mat_List]:=
 Module[{i = 2, clusterCornerID, cornerLabels, reLabel, Moore},
  clusterCornerID[1, 0, 0, 0] := i++;
  clusterCornerID[a_, __] := a;
  Attributes[clusterCornerID] = Listable;

  cornerLabels = clusterCornerID[#, RotateRight[#, {1, 0}],
                        RotateRight[#, {0, 1}],
                        RotateRight[#, {1, 1}]]&[mat];

  reLabel[0, ___] := 0;
  reLabel[a_, b_, c_, d_, e_, f_, g_, h_, i_] :=
                        Max[a, b, c, d, e, f, g, h, i];
```

```
Moore[func__, lat_] :=
  MapThread[func, Map[RotateRight[lat, #]&,
            {{0, 0}, {1, 0}, {0, -1}, {-1, 0}, {0, 1},
             {1, -1}, {-1, -1}, {-1, 1}, {1, 1}}], 2];

  FixedPoint[Moore[reLabel, #]&, cornerLabels]
  ]
```

The `clusterLabelMoore` program labels clusters in the same way as the `clusterLabelvonNeumann` program, starting in the bottom row. Therefore, in order to use the program to identify the empty site clusters in the sapoval lattice, we change the 0's into 1's and 1's into 0's and reverse the order of the rows in the lattice.

```
BClustersLat = clusterLabelMoore[Reverse[sapoval /. {0 -> 1, 1 -> 0}]]
```

Taking BClustersLat, reversing the rows and renumbering the clusters of empty sites, we get

```
BClusters =
 Reverse[BClustersLat /.
  MapThread[Rule,
           {Reverse[Rest[Union[Flatten[BClustersLat]]]],
            Range[Length[Union[Flatten[BClustersLat]]] - 1]}]
           ]
```

```
{{1, 1, 1, 1, 1, 1, 1, 1, 1, 1, 1, 1, 1, 1, 1, 1, 1},
 {1, 0, 0, 1, 1, 1, 1, 1, 1, 1, 1, 1, 1, 1, 1, 1, 1},
 {1, 1, 1, 1, 1, 1, 1, 1, 1, 1, 0, 1, 1, 1, 1, 1, 1},
 {1, 1, 0, 1, 1, 0, 1, 1, 0, 1, 0, 0, 1, 1, 1, 1, 1},
 {1, 1, 0, 1, 1, 0, 1, 1, 0, 1, 1, 1, 1, 0, 1, 1, 1},
 {0, 1, 1, 1, 0, 1, 1, 1, 1, 0, 0, 1, 1, 1, 1, 0, 1},
 {1, 1, 1, 0, 0, 1, 1, 1, 0, 0, 0, 1, 0, 1, 1, 1, 1},
 {1, 0, 1, 1, 0, 1, 1, 0, 1, 1, 1, 1, 0, 1, 1, 1, 1},
 {0, 0, 0, 1, 1, 1, 1, 1, 0, 0, 1, 0, 0, 1, 0, 1, 1},
 {0, 0, 0, 0, 0, 0, 0, 0, 1, 0, 1, 1, 1, 0, 0, 1, 0},
 {0, 0, 0, 0, 1, 1, 1, 1, 0, 0, 1, 0, 1, 0, 0, 0, 0},
 {0, 0, 0, 0, 1, 0, 0, 1, 0, 0, 0, 0, 1, 0, 0, 0, 2},
 {0, 0, 0, 0, 1, 1, 0, 0, 1, 0, 0, 0, 0, 0, 3, 0, 2},
 {0, 0, 0, 0, 0, 0, 0, 0, 0, 0, 4, 0, 0, 0, 0, 0, 0},
 {0, 6, 6, 0, 0, 0, 0, 0, 5, 0, 0, 0, 0, 0, 0, 0, 0},
 {0, 0, 0, 0, 0, 0, 0, 0, 0, 0, 0, 0, 0, 0, 0, 0, 0},
 {0, 0, 0, 0, 0, 0, 0, 0, 0, 0, 0, 0, 0, 0, 0, 0, 0}}
```

The sites having value 1 in BClusters belong to the infinite empty site cluster in the sapoval lattice. Their locations in the lattice is given by

```
infiniteBClusterlocs = Position[BClusters, 1]
```

```
{{1, 1}, {1, 2}, {1, 3}, {1, 4}, {1, 5}, {1, 6}, {1, 7},
 {1, 8}, {1, 9}, {1, 10}, {1, 11}, {1, 12}, {1, 13}, {1, 14},
 {1, 15}, {1, 16}, {1, 17}, {2, 1}, {2, 4}, {2, 5}, {2, 6},
 {2, 7}, {2, 8}, {2, 9}, {2, 10}, {2, 11}, {2, 12}, {2, 13},
 {2, 14}, {2, 15}, {2, 16}, {2, 17}, {3, 1}, {3, 2}, {3, 3},
 {3, 4}, {3, 5}, {3, 6}, {3, 7}, {3, 8}, {3, 9}, {3, 10},
 {3, 12}, {3, 13}, {3, 14}, {3, 15}, {3, 16}, {3, 17},
 {4, 1}, {4, 2}, {4, 4}, {4, 5}, {4, 7}, {4, 8}, {4, 10},
 {4, 13}, {4, 14}, {4, 15}, {4, 16}, {4, 17}, {5, 1}, {5, 2},
 {5, 4}, {5, 5}, {5, 7}, {5, 8}, {5, 10}, {5, 11}, {5, 12},
 {5, 13}, {5, 15}, {5, 16}, {5, 17}, {6, 2}, {6, 3}, {6, 4},
 {6, 6}, {6, 7}, {6, 8}, {6, 9}, {6, 12}, {6, 13}, {6, 14},
 {6, 15}, {6, 17}, {7, 1}, {7, 2}, {7, 3}, {7, 6}, {7, 7},
 {7, 8}, {7, 12}, {7, 14}, {7, 15}, {7, 16}, {7, 17}, {8, 1},
 {8, 3}, {8, 4}, {8, 6}, {8, 7}, {8, 9}, {8, 10}, {8, 11},
 {8, 12}, {8, 14}, {8, 15}, {8, 16}, {8, 17}, {9, 4}, {9, 5},
 {9, 6}, {9, 7}, {9, 8}, {9, 11}, {9, 14}, {9, 16}, {9, 17},
 {10, 9}, {10, 11}, {10, 12}, {10, 13}, {10, 16}, {11, 5},
 {11, 6}, {11, 7}, {11, 8}, {11, 11}, {11, 13}, {12, 5},
 {12, 8}, {12, 13}, {13, 5}, {13, 6}, {13, 9}}}
```

Now that we have the list, infiniteAClusterlocs, of the sites in the infinite occupied site cluster, and the list, infiniteBClusterlocs, of the sites in the infinite empty site cluster, it is straightforward to determine the list of sites, frontierBSites, in infiniteBClusterlocs that lie in the von Neumann neighborhood of sites in infiniteAClusterlocs, and the list of sites, frontierASites, in infiniteAClusterlocs that lie in the Moore neighborhood of sites in infiniteBClusterlocs.

```
frontierBSites =
DeleteCases[
 Select[infiniteBClusterlocs,
   MemberQ[
    (Flatten[
    Map[Function[y, Map[(y + #)&,{{1,0},{-1,0},{0,1},{0,-1}}]],
      infiniteAClusterlocs], 1] /.
       {0 ->17, 18 -> 1}), #]&], {1, _}]
```

```
{{7, 2}, {8, 1}, {8, 3}, {8, 9}, {8, 10}, {8, 15}, {9, 4},
 {9, 5}, {9, 6}, {9, 7}, {9, 8}, {9, 11}, {9, 14}, {9, 16},
 {9, 17}, {10, 9}, {10, 11}, {10, 12}, {10, 13}, {10, 16},
 {11, 5}, {11, 6}, {11, 7}, {11, 8}, {11, 11}, {11, 13},
 {12, 5}, {12, 8}, {12, 13}, {13, 5}, {13, 6}, {13, 9}}
```

```
frontierASites =
DeleteCases[
 Select[infiniteAClusterlocs,
  MemberQ[
  (Flatten[
  Map[Function[y,
    Map[(y + #)&, {{1,0},{-1,0},{0,1},{0,-1},
                   {1,1},{1,-1},{-1,1},{-1,-1}}]],
     infiniteBClusterlocs], 1] /.
       {0 ->17, 18 -> 1}), #]&], {17, _}]
```

```
{{8, 2}, {9, 1}, {9, 2}, {9, 3}, {9, 9}, {9, 10}, {9, 15},
 {10, 1}, {10, 3}, {10, 4}, {10, 5}, {10, 6}, {10, 7},
 {10, 8}, {10, 10}, {10, 14}, {10, 15}, {10, 17}, {11, 4},
 {11, 9}, {11, 10}, {11, 12}, {11, 14}, {11, 15}, {11, 16},
 {11, 17}, {12, 4}, {12, 6}, {12, 7}, {12, 9}, {12, 10},
 {12, 11}, {12, 12}, {12, 14}, {13, 4}, {13, 7}, {13, 8},
 {13, 10}, {13, 12}, {13, 13}, {13, 14}, {14, 4}, {14, 5},
 {14, 6}, {14, 7}, {14, 8}, {14, 9}, {14, 10}}
```

Using the sapoval lattice and the lists of the sites along the two diffusion fronts, frontierASites and frontierBSites, we can create a graphic in which the A sites are colored blue, the A frontier sites are colored green, and the B frontier sites are colored red.

```
Show[Graphics[RasterArray[
 Reverse[ReplacePart[ReplacePart[sapoval, FA, frontierASites],
              FB, frontierBSites]] /.
              {0 -> RGBColor[1, 1, 1], 1 -> RGBColor[0, 0, 1],
                FA -> RGBColor[0, 1, 0], FB -> RGBColor[1, 0, 0]}],
   AspectRatio -> Automatic, Frame -> True,
   FrameTicks -> None]];
```

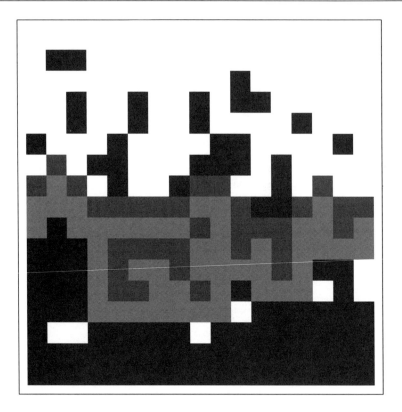

Computer Simulation Projects

1. Create a program for calculating the diffusion fronts of a two-dimensional gradient percolation system, using the code fragments presented here.

2. An analytical analysis indicates that the concentration $p[x]$ follows so-called Fickian behavior, going as $\text{Erfc}[x/L]$ where x is the distance from the source and $L = 2(D * t)^{0.5}$ is the diffusion length at time t with diffusion coefficient D. Using this concentration profile, determine the power law dependencies of the number of particles on the front and the spread (width) of the front on the diffusion length.

3. A three-dimensional gradient percolation system (Rosso, et al., 1986) can be created using

```
gradientPercolation3D[n_, m_] :=
 Module[{p = 0},

  laydown =
   (p += 1/(n - 1); Append[#, Table[Floor[Random[] + p], {m},{m}]] )&;

  Nest[laydown,{Table[0, {m}, {m}]}, n - 1]
  ];
```

A graphic of the occupied sites in the system can be created by locating the positions of sites with value 1 in the lattice and applying the Cuboid graphic primitive to those positions.

For example,

```
Show[Graphics3D[
             Map[Cuboid, Position[gradientPercolation3D[9, 10], 1]]],
       ViewPoint->{-0.995, 1.935, 2.591}];
```

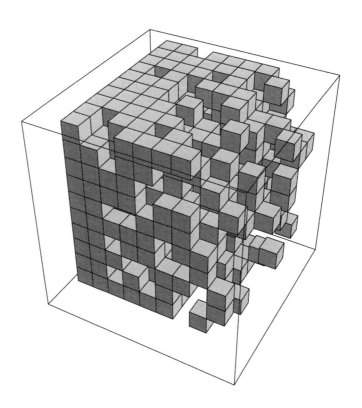

Defining the interface so that the boundaries of the two infinite clusters are in contact with each other everywhere, each site in the cluster of occupied sites is one of the four nearest neighbor sites in the N, E, S, W directions of another site in the cluster, while each site in the cluster of empty sites is one of the 12 sites that are either nearest neighbor sites in the N, NE, E, SE, S, SW, W, NW directions, or next nearest neighbor sites in the N, E, S, W directions of another site in the cluster.

References

Chopard, Bastien, Droz, Michael, and Kolb, Max. 1989. "Cellular automata approach to non-equilibrium diffusion and gradient percolation." J. Phys. A, 22, 1609–1619.

Rosso, M., Gouyet, J. F., and Sapoval, B. 1986. "Gradient percolation in three dimensions and relation to diffusion fronts." Phys. Rev. Lett., 57, 3195–3198.

Sapoval, B., Rosso, M., and Gouyet, J. F. 1985. "The fractal nature of a diffusion front and the relation to percolation." J. Physique Lett., 46, L149–L156.

9 | Two-Species Driven Diffusion

Introduction

The random motion of the particles in a two-species system in the presence of an external field, originally proposed as a model of certain kinds of materials known as superionic conductors, fast ion conductors, super ionic solids, or solid electrolytes shows rather interesting steady-state, nonequilibrium behavior. In this *driven diffusive* system, the external field biases the movement of each species in opposite directions and an order-disorder transition occurs as a function of density. We will develop a CA model of the driven diffusion of two species.

The Two-Species Driven Diffusion CA

The System

The CA employs an *n*-by-*n* square lattice with periodic boundary conditions. Lattice sites are either empty or occupied by a single particle that has either a positive or negative charge. The total density of occupied sites is p, and there are (approximately) the same number of positive and negative particles so that total charge of the system is (approximately) zero. Initially, the particles are randomly distributed on the lattice and each particle faces a randomly chosen direction. The system is created using the following code:

```
initPhase =
    Table[{{1, 2, 3, 4}[[Random[Integer, {1, 4}]]],
            {A, B}[[Random[Integer, {1, 2}]]]} * Floor[Random[] + p],
        {n}, {n}];
```

The sites in initPhase have values that are ordered pairs {x, y}. The first component of each ordered pair is an integer ranging from 0 to 4 where the values have the following meanings:

0—an empty site
1—a site occupied by a north-facing particle
2—a site occupied by an east-facing particle
3—a site occupied by a south-facing particle
4—a site occupied by a west-facing particle

The second component of each ordered pair is either 0, A, or B, where 0 indicates an empty site, A indicates a site occupied by an A type particle, and B indicates a site occupied by a B type particle.

The CA Rules

The rules for the behavior of a particle during a time step take 13 arguments:

```
driven[site, N, E, S, W, NE, SE, SW, NW, Nn, Ee, Ss, Ww]
```

The arguments represent the value of the site and the values of the nearest neighbors in the N, E, S, W directions, the nearest neighbors in the NE, SE, SW, NW directions, and the next nearest neighbors in the N, E, S, W directions (these values are needed to implement the "excluded volume" constraint).

The presence of an external electrical field causes an A (B) particle to hop preferentially along (against) the field direction. The field points diagonally in the north-east direction, which drives the A (B) particles upward (downward). The probability of a step is a in the preferred direction (north and east for an A particle, south and west for a B particle) and $(1 - a)$ in the opposite direction.

In the CA, a particle will always move in the direction it is facing, and therefore we will implement the bias in taking a step in various directions by biasing the probabilities that a particle chooses to face those directions. An A particle will choose a direction to face using rndA, and a B particle will choose a direction to face using rndB where rndA and rndB are defined as follows:

```
rndA := {Random[Integer, {1, 2}],
         Random[Integer, {3, 4}]}[[2 -Floor[Random[] + a]]]
```

With rndA, 1, 2, 3, and 4 are chosen $(a/2)\%$, $(a/2)\%$, $(1-a)/2\%$, and $(1-a)/2\%$ of the time, respectively.

```
rndB := {Random[Integer, {1, 2}],
         Random[Integer, {3, 4}]}[[1 + Floor[Random[] + a]]]
```

With rndB, 1, 2, 3, and 4 are chosen $(1-a)/2\%$, $(1-a)/2\%$, $(a/2)\%$, and $(a/2)\%$ of the time, respectively.

At each time step, all of the particles simultaneously attempt to jump to adjacent sites. When a particle attempts to move to the nearest neighbor site it faces, there are three possible situations that will determine the outcome:

(a) The adjacent site is occupied by another particle, in which case the particle remains in place and randomly chooses a direction to face.

(b) The adjacent site is empty and is faced by one or more other particles, in which case the particle remains in place and randomly chooses a direction to face.

(c) The adjacent site is empty and no other particles face it, in which case the particle abandons its current site and moves into the site and randomly chooses a direction to face.

Regardless of whether or not a particle is able to move, it randomly picks a direction to face using rndA if it is an A particle and using rndB if it is a B particle.

The following rules describe the behavior of lattice sites during a time step:

- A particle facing a site occupied by another particle (of either type) remains in place and randomly chooses a direction to face.

```
driven[{1, A}, {x_?Positive, _}, _, _, _, _, _, _, _, _, _, _, _] :=
                                                          {rndA, A}
driven[{2, A}, _, {x_?Positive, _}, _, _, _, _, _, _, _, _, _] :=
                                                          {rndA, A}
driven[{3, A}, _, _, {x_?Positive, _}, _, _, _, _, _, _, _, _] :=
                                                          {rndA, A}
driven[{4, A}, _, _, _, {x_?Positive, _}, _, _, _, _, _, _, _] :=
                                                          {rndA, A}
driven[{1, B}, {x_?Positive, _}, _, _, _, _, _, _, _, _, _, _] :=
                                                          {rndB, B}
driven[{2, B}, _, {x_?Positive, _}, _, _, _, _, _, _, _, _, _] :=
                                                          {rndB, B}
```

```
driven[{3, B}, _, _, {x_?Positive, _}, _, _, _, _, _, _, _, _, _] :=
                                                             {rndB, B}
driven[{4, B}, _, _, _, {x_?Positive, _}, _, _, _, _, _, _, _, _] :=
                                                             {rndB, B}
```

- A particle facing an empty adjacent site that is also faced by another
  particle of either type (e.g., a right-facing particle that lies to the left of
  an empty site that is faced by a left-facing particle lying to the right of
  the empty site) stays put and randomly chooses a direction to face.

```
driven[{1, A}, {0, 0}, _, _, _, {4, _}, _, _, _, _, _, _, _] :=
                                                             {rndA, A}
driven[{1, A}, {0, 0}, _, _, _, _, _, _, {2, _}, _, _, _, _] :=
                                                             {rndA, A}
driven[{1, A}, {0, 0}, _, _,_ ,_ ,_ ,_ , _, {3, _}, _, _, _] :=
                                                             {rndA, A}
driven[{2, A}, _, {0, 0}, _, _, {3, _}, _, _, _, _, _, _, _] :=
                                                             {rndA, A}
driven[{2, A}, _, {0, 0}, _, _, _, {1, _}, _, _, _, _, _, _] :=
                                                             {rndA, A}
driven[{2, A}, _, {0, 0}, _, _, _, _, _, _, _, {4, _}, _, _] :=
                                                             {rndA, A}
driven[{3, A}, _, _, {0, 0}, _, _, {4, _}, _, _, _, _, _, _] :=
                                                             {rndA, A}
driven[{3, A}, _, _, {0, 0}, _, _, _, {2, _}, _, _, _, _, _] :=
                                                             {rndA, A}
driven[{3, A}, _, _, {0, 0}, _, _, _, _, _, _, _, {1, _}, _] :=
                                                             {rndA, A}
driven[{4, A}, _, _, _, {0, 0}, _, _, {1, _}, _, _, _, _, _] :=
                                                             {rndA, A}
driven[{4, A}, _, _, _, {0, 0}, _, _, _, {3, _}, _, _, _, _] :=
                                                             {rndA, A}
driven[{4, A}, _, _, _, {0, 0}, _, _, _, _, _, _, _, {2, _}] :=
                                                             {rndA, A}
driven[{1, B}, {0, 0}, _, _, _, {4, _}, _, _, _, _, _, _, _] :=
                                                             {rndB, B}
driven[{1, B}, {0, 0}, _, _, _, _, _, _, {2, _}, _, _, _, _] :=
                                                             {rndB, B}
driven[{1, B}, {0, 0}, _, _,_ ,_ ,_ ,_ , _, {3, _}, _, _, _] :=
                                                             {rndB, B}
driven[{2, B}, _, {0, 0}, _, _, {3, _}, _, _, _, _, _, _, _] :=
                                                             {rndB, B}
driven[{2, B}, _, {0, 0}, _, _, _, {1, _}, _, _, _, _, _, _] :=
                                                             {rndB, B}
```

```
driven[{2, B}, _, {0, 0}, _, _, _, _, _, _, _, {4, _}, _, _] :=
                                                      {rndB, B}
driven[{3, B}, _, _, {0, 0}, _, _, {4, _}, _, _, _, _, _, _] :=
                                                      {rndB, B}
driven[{3, B}, _, _, {0, 0}, _, _, _, {2, _}, _, _, _, _, _] :=
                                                      {rndB, B}
driven[{3, B}, _, _, {0, 0}, _, _, _, _, _, _, _, {1, _}, _] :=
                                                      {rndB, B}
driven[{4, B}, _, _, _, {0, 0}, _, _, {1, _}, _, _, _, _, _] :=
                                                      {rndB, B}
driven[{4, B}, _, _, _, {0, 0}, _, _, _, {3, _}, _, _, _, _] :=
                                                      {rndB, B}
driven[{4, B}, _, _, _, {0, 0}, _, _, _, _, _, _, _, {2, _}] :=
                                                      {rndB, B}
```

- A site occupied by a particle facing an empty adjacent site that is not faced by another particle is vacated.

```
driven[{1, _}, {0, 0}, _, _, _, _, _, _, _, _, _, _] := {0, 0}
driven[{2, _}, _, {0, 0}, _, _, _, _, _, _, _, _, _] := {0, 0}
driven[{3, _}, _, _, {0, 0}, _, _, _, _, _, _, _, _] := {0, 0}
driven[{4, _}, _, _, _, {0, 0}, _, _, _, _, _, _, _] := {0, 0}
```

- An empty site faced by two or more particles remains empty.

```
driven[{0, 0}, {3, _}, {4, _}, _, _, _, _, _, _, _, _, _] := {0, 0}
driven[{0, 0}, {3, _}, _, {1, _}, _, _, _, _, _, _, _, _] := {0, 0}
driven[{0, 0}, {3, _}, _, _, {2, _}, _, _, _, _, _, _, _] := {0, 0}
driven[{0, 0}, _, {4, _}, {1, _}, _, _, _, _, _, _, _, _] := {0, 0}
driven[{0, 0}, _, {4, _}, _, {2, _}, _, _, _, _, _, _, _] := {0, 0}
driven[{0, 0}, _, _, {1, _}, {2, _}, _, _, _, _, _, _, _] := {0, 0}
```

- An empty site faced by exactly one particle becomes occupied.

```
driven[{0, 0}, {3, A}, _, _, _, _, _, _, _, _, _, _] := {rndA, A}
driven[{0, 0}, _, {4, A}, _, _, _, _, _, _, _, _, _] := {rndA, A}
driven[{0, 0}, _, _, {1, A}, _, _, _, _, _, _, _, _] := {rndA, A}
driven[{0, 0}, _, _, _, {2, A}, _, _, _, _, _, _, _] := {rndA, A}
driven[{0, 0}, {3, B}, _, _, _, _, _, _, _, _, _, _] := {rndB, B}
driven[{0, 0}, _, {4, B}, _, _, _, _, _, _, _, _, _] := {rndB, B}
driven[{0, 0}, _, _, {1, B}, _, _, _, _, _, _, _, _] := {rndB, B}
driven[{0, 0}, _, _, _, {2, B}, _, _, _, _, _, _, _] := {rndB, B}
```

- An empty site remains unchanged.

```
driven[{0, 0}, _, _, _, _, _, _, _, _, _, _, _, _] := {0, 0}
```

Note: The number of rules (51) given above is rather large and can be shortened to 31 (with some loss of readability and a small decrease in efficiency) by making the following changes:

(a) using 1 and 2 as the second component of the ordered pair value of a lattice site, rather than A and B, to indicate the type of particle.

(b) replacing rndA and rndB with

```
rnd := {{Random[Integer, {1, 2}],
         Random[Integer, {3, 4}]}[[2 - Floor[Random[] + a]]],
        {Random[Integer, {1, 2}],
         Random[Integer, {3, 4}]}[[1 + Floor[Random[] + a]]]}
```

(c) combining the rules that are the same except for the type of particle they apply to. For example, the two rules

```
driven[{1, A}, {x_?Positive, _}, _, _, _, _, _, _, _, _, _, _, _] :=
                                                            {rndA, A}
driven[{1, B}, {x_?Positive, _}, _, _, _, _, _, _, _, _, _, _, _] :=
                                                            {rndB, B}
```

can be merged into the single rule

```
driven[{1, y_}, {x_?Positive, _}, _, _, _, _, _, _, _, _, _, _, _] :=
                                                         {rnd[[y]], y}
```

Applying the CA Rules

The update rules are applied to the CA lattice using the anonymous function

```
MvonN[driven, #]&
```

where

```
MvonN[func__, lat_] :=
  MapThread[func, Map[RotateRight[lat, #]&,
            {{0, 0}, {1, 0}, {0, -1}, {-1, 0}, {0, 1},
             {1, -1}, {-1, -1}, {-1, 1}, {1, 1},
             {2, 0}, {0, -2}, {-2, 0}, {0, 2}}], 2]
```

The positions of the random walkers evolve over *t* time steps, starting with the initial lattice configuration, using the following NestList operation.

```
NestList[MvonN[driven, #]&, initPhase, t]
```

The Program

```
DrivenDiffusion2Species[n_, p_, a_, t_]:=
Module[{initPhase, driven, rndA, rndB},

initPhase =
  Table[{{1, 2, 3, 4}[[Random[Integer, {1, 4}]]],
         {A, B}[[Random[Integer, {1, 2}]]]} * Floor[Random[] + p],
        {n}, {n}];

rndA := {Random[Integer, {1, 2}],
         Random[Integer, {3, 4}]}[[2 -Floor[Random[] + a]]];

rndB := {Random[Integer, {1, 2}],
         Random[Integer, {3, 4}]}[[1 + Floor[Random[] + a]]];

driven[{1, A}, {x_?Positive, _}, _, _, _, _, _, _, _, _, _, _, _] :=
                                                   {rndA, A};
driven[{2, A}, _, {x_?Positive, _}, _, _, _, _, _, _, _, _, _, _] :=
                                                   {rndA, A};
driven[{3, A}, _, _, {x_?Positive, _}, _, _, _, _, _, _, _, _, _] :=
                                                   {rndA, A};
driven[{4, A}, _, _, _, {x_?Positive, _}, _, _, _, _, _, _, _, _] :=
                                                   {rndA, A};
driven[{1, B}, {x_?Positive, _}, _, _, _, _, _, _, _, _, _, _, _] :=
                                                   {rndB, B};
driven[{2, B}, _, {x_?Positive, _}, _, _, _, _, _, _, _, _, _, _] :=
                                                   {rndB, B};
driven[{3, B}, _, _, {x_?Positive, _}, _, _, _, _, _, _, _, _, _] :=
                                                   {rndB, B};
driven[{4, B}, _, _, _, {x_?Positive, _}, _, _, _, _, _, _, _, _] :=
                                                   {rndB, B};
driven[{1, A}, {0, 0}, _, _, _, {4, _}, _, _, _, _, _, _, _] :=
                                                   {rndA, A};
driven[{1, A}, {0, 0}, _, _, _, _, _, _, {2, _}, _, _, _, _] :=
                                                   {rndA, A};
driven[{1, A}, {0, 0}, _, _,_ ,_ ,_ ,_ , _, {3, _}, _, _, _] :=
                                                   {rndA, A};
```

```
driven[{2, A}, _, {0, 0}, _, _, {3, _}, _, _, _, _, _, _, _] :=
                                              {rndA, A};
driven[{2, A}, _, {0, 0}, _, _, _, {1, _}, _, _, _, _, _, _] :=
                                              {rndA, A};
driven[{2, A}, _, {0, 0}, _, _, _, _, _, _, _, {4, _}, _, _] :=
                                              {rndA, A};
driven[{3, A}, _, _, {0, 0}, _, _, {4, _}, _, _, _, _, _, _] :=
                                              {rndA, A};
driven[{3, A}, _, _, {0, 0}, _, _, _, {2, _}, _, _, _, _, _] :=
                                              {rndA, A};
driven[{3, A}, _, _, {0, 0}, _, _, _, _, _, _, _, {1, _}, _] :=
                                              {rndA, A};
driven[{4, A}, _, _, _, {0, 0}, _, _, {1, _}, _, _, _, _, _] :=
                                              {rndA, A};
driven[{4, A}, _, _, _, {0, 0}, _, _, _, {3, _}, _, _, _, _] :=
                                              {rndA, A};
driven[{4, A}, _, _, _, {0, 0}, _, _, _, _, _, _, _, {2, _}] :=
                                              {rndA, A};
driven[{1, B}, {0, 0}, _, _, _, {4, _}, _, _, _, _, _, _, _] :=
                                              {rndB, B};
driven[{1, B}, {0, 0}, _, _, _, _, _, _, {2, _}, _, _, _, _] :=
                                              {rndB, B};
driven[{1, B}, {0, 0}, _, _, _, _, _, _, _, {3, _}, _, _, _] :=
                                              {rndB, B};
driven[{2, B}, _, {0, 0}, _, _, {3, _}, _, _, _, _, _, _, _] :=
                                              {rndB, B};
driven[{2, B}, _, {0, 0}, _, _, _, {1, _}, _, _, _, _, _, _] :=
                                              {rndB, B};
driven[{2, B}, _, {0, 0}, _, _, _, _, _, _, _, {4, _}, _, _] :=
                                              {rndB, B};
driven[{3, B}, _, _, {0, 0}, _, _, {4, _}, _, _, _, _, _, _] :=
                                              {rndB, B};
driven[{3, B}, _, _, {0, 0}, _, _, _, {2, _}, _, _, _, _, _] :=
                                              {rndB, B};
driven[{3, B}, _, _, {0, 0}, _, _, _, _, _, _, _, {1, _}, _] :=
                                              {rndB, B};
driven[{4, B}, _, _, _, {0, 0}, _, _, {1, _}, _, _, _, _, _] :=
                                              {rndB, B};
driven[{4, B}, _, _, _, {0, 0}, _, _, _, {3, _}, _, _, _, _] :=
                                              {rndB, B};
driven[{4, B}, _, _, _, {0, 0}, _, _, _, _, _, _, _, {2, _}] :=
                                              {rndB, B};
driven[{1, _}, {0, 0}, _, _, _, _, _, _, _, _, _, _, _] := {0, 0};
driven[{2, _}, _, {0, 0}, _, _, _, _, _, _, _, _, _, _] := {0, 0};
driven[{3, _}, _, _, {0, 0}, _, _, _, _, _, _, _, _, _] := {0, 0};
```

```
driven[{4, _}, _, _, _, {0, 0}, _, _, _, _, _, _, _, _] := {0, 0};
driven[{0, 0}, {3, _}, {4, _}, _, _, _, _, _, _, _, _, _]:={0, 0};
driven[{0, 0}, {3, _}, _, {1, _}, _, _, _, _, _, _, _, _]:={0, 0};
driven[{0, 0}, {3, _}, _, _, {2, _}, _, _, _, _, _, _, _]:={0, 0};
driven[{0, 0}, _, {4, _}, {1, _}, _, _, _, _, _, _, _, _]:={0, 0};
driven[{0, 0}, _, {4, _}, _, {2, _}, _, _, _, _, _, _, _]:={0, 0};
driven[{0, 0}, _, _, {1, _}, {2, _}, _, _, _, _, _, _, _]:={0, 0};
driven[{0, 0}, {3, A}, _, _, _, _, _, _, _, _, _, _] := {rndA, A};
driven[{0, 0}, _, {4, A}, _, _, _, _, _, _, _, _, _] := {rndA, A};
driven[{0, 0}, _, _, {1, A}, _, _, _, _, _, _, _, _] := {rndA, A};
driven[{0, 0}, _, _, _, {2, A}, _, _, _, _, _, _, _] := {rndA, A};
driven[{0, 0}, {3, B}, _, _, _, _, _, _, _, _, _, _] := {rndB, B};
driven[{0, 0}, _, {4, B}, _, _, _, _, _, _, _, _, _] := {rndB, B};
driven[{0, 0}, _, _, {1, B}, _, _, _, _, _, _, _, _] := {rndB, B};
driven[{0, 0}, _, _, _, {2, B}, _, _, _, _, _, _, _] := {rndB, B};
driven[{0, 0}, _, _, _, _, _, _, _, _, _, _, _] := {0, 0};


MvonN[func__, lat_] :=
  MapThread[func, Map[RotateRight[lat, #]&,
               {{0, 0}, {1, 0}, {0, -1}, {-1, 0}, {0, 1},
                {1, -1}, {-1, -1}, {-1, 1}, {1, 1},
                {2, 0}, {0, -2}, {-2, 0}, {0, 2}}], 2];


NestList[MvonN[driven, #]&, initPhase, t]}
]
```

Running the Program

The lattice configuration at each time step, showing whether a site is empty or is occupied by an A particle or a B particle can be determined by extracting the second component of the ordered pair value of each lattice site, using

```
DrivenDiffusion2Species[n, p, a, t] /.
                {{_, 0} -> 0, {_, A} -> A, {_, B} -> B}
```

For example, we can look at the development over 100 time steps, of a 30-by-30 lattice system with particles having a 90% probability of moving in the preferred direction and a 10% probability of moving in the opposite direction. We can compare the low density ($p = 0.1$) and high density ($p = 0.5$) cases.

```
lowDensityCase = DrivenDiffusion2Species[30, 0.1, 0.9, 100] /.
                       {{_, 0} -> 0, {_, A} -> A, {_, B} -> B};


Show[GraphicsArray[Partition[
  Map[Show[Graphics[RasterArray[# /. {0 -> RGBColor[0.7, 0.7, 0.7],
                                      A -> RGBColor[0, 1, 0],
                                      B -> RGBColor[1, 0, 1]}]],
           AspectRatio -> Automatic, DisplayFunction -> Identity]&,
      lowDensityCase[[Range[1, 101, 20] ]]], 2]]];
```

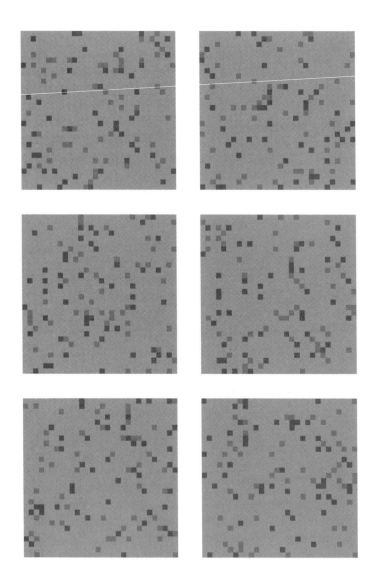

```
highDensityCase = DrivenDiffusion2Species[30, 0.5, 0.9, 100] /.
                         {{_, 0} -> 0, {_, A} -> A, {_, B} -> B};

Show[GraphicsArray[Partition[Map[Show[Graphics[RasterArray[# /.
          {0 -> RGBColor[0.7, 0.7, 0.7], A -> RGBColor[0, 1, 0],
           B -> RGBColor[1, 0, 1]}]]],
          AspectRatio -> Automatic, DisplayFunction -> Identity
               ]&,
          highDensityCase[[Range[1, 101, 20] ]]
     ], 2] ]];
```

At low densities, the lattice configuration remains disordered over time with a homogeneous distribution of the holes and the A and B particles. The current in this system is relatively high.

As the density of particles increases, the particles mutually block and impede one another, giving rise to inhomogeneities in the distribution of the holes and the A and B particles, and in the charge density and mass density (particle density regardless of charge) distributions. Eventually, there is a transition resulting in phase segregation and the emergence of a striped, banded structure. The space between the strips consists primarily of holes. Within each strip, there are two substrips of approximately equal size, one comprised of A particles and one comprised of B particles. The "locking up" of the system by the "mutual blocking" effect significantly reduces the current in the system.

Note: The transition in the distribution of particles (and hence of charge) as a function of particle density is analogous to the "jamming" transition seen in another driven diffusive system, traffic flow (see Gaylord and Wellin, 1995).

Computer Simulation Project

We can locate the critical value at which the first-order phase transition in the two-species driven diffusive system occurs by visually identifying the lowest density at which an inhomogeneous banded structure appears in the lattice. It should also be possible to use some quantitative measure whose value changes abruptly at the transition point (e.g., the density profile of the holes is quite different in the homogeneous and inhomogeneous states). Think of some quantity that can be computed from the lattice configuration, and create a plot of that quantity as a function of density to locate the critical density at which the value of this quantity changes abruptly.

References

Gaylord, Richard J. and Wellin, Paul R. *Computer Simulations with Mathematica: Explorations in Complex Physical and Biological Systems*. TELOS/Springer-Verlag (1995).

Kertesz, J.., and Ramawamy, R. 1994. "Coarsening in a driven diffusive system with two species." Europhysics Letters, 28, 617–622.

Schmittmann, B., Hwang, K., and Zia, R.K.P. 1992. "Onset of spatial structures in biased diffusion of two species." Europhysics Letters, 19, 19–25.

Schmittmann, B. and Zia, R.K.P. *Statistical Mechanics of Driven Diffusive Systems*. Academic Press (1995).

10 Coalescence

Introduction

Coalescence refers to the growing or joining together of objects into a single body. The phenomenon of coalescence has both technological and scientific importance, occurring in a variety of systems, including aerosols (e.g., clouds, fog, atmospheric pollution), condensed materials (e.g., vapor condensation on surfaces, droplet growth on vapor deposited films, step-growth polymerization), and societies (in ethology, *biosocial attraction* refers to the tendency of animals, such as fish, to group together).

We will first present a cellular automaton model of the coalescence of diffusing particles. Then, after showing how to calculate the distribution of particle sizes, we will discuss how to implement various mechanisms, such as the injection of small particles or the removal of large particles, which result in a "steady-state" distribution of particle sizes (which commonly follow power law behavior).

The Coalescence CA

The System

The system consists of an n-by-n square lattice with periodic boundary conditions. The values of the lattice sites are ordered pairs $\{x, y\}$.

The first component of the ordered pair is an integer ranging from 0 to 4 where the values have the following meanings:

0 — an empty site
1 — a site occupied by a north-facing droplet
2 — a site occupied by an east-facing droplet
3 — a site occupied by a south-facing droplet
4 — a site occupied by a west-facing droplet

The second component of the ordered pair is a nonnegative integer that represents the size of the droplet on the site.

The initial CA lattice configuration, called "cloud," consists of randomly distributed droplets of size 1, so that the lattice sites are either empty or occupied by a droplet of size 1, with a probability p.

```
cloud = Table[{RND, 1} * Floor[Random[] + p], {n}, {n}]
```

where

```
RND := Random[Integer, {1, 4}]}
```

Note: The density of droplets in the cloud approaches p as the system size increases.

Moving and Coalescing Droplets

In a given time step, all of the droplets, without exception, move to the adjacent site they are facing. When more than one droplet moves to the same site, they merge into a single droplet. This process is carried out in the CA using 16 rules having the form

```
moveCoalesce[site, N, E, S, W]
```

where moveCoalesce takes the values of a site and its four nearest neighbors as arguments.

Fifteen of the moveCoalesce rules take droplets that face a common nearest neighbor site, move them to the site, merge them into a single droplet (whose size is the sum of the sizes of the particles moving to the site), and face the resulting droplet in a randomly chosen direction.

```
moveCoalesce[_, {3, a_}, {4, b_}, {1, c_}, {2, d_}] :=
                                              {RND, a + b + c + d}

moveCoalesce[_, {3, a_}, _, {1, c_}, {2, d_}] := {RND, a + c + d}
moveCoalesce[_, _, {4, b_}, {1, c_}, {2, d_}] := {RND, b + c + d}
moveCoalesce[_, {3, a_}, {4, b_}, {1, c_}, _] := {RND, a + b + c}
moveCoalesce[_, {3, a_}, {4, b_}, _, {2, d_}] := {RND, a + b + d}

moveCoalesce[_, _, _, {1, c_}, {2, d_}] := {RND, c + d}
moveCoalesce[_, {3, a_}, _, {1, c_}, _] := {RND, a + c}
moveCoalesce[_, _, {4, b_}, {1, c_}, _] := {RND, b + c}
moveCoalesce[_, {3, a_}, _, _, {2, d_}] := {RND, a + d}
```

```
moveCoalesce[_, _, {4, b_}, _, {2, d_}] := {RND, b + d}
moveCoalesce[_, {3, a_}, {4, b_}, _, _] := {RND, a + b}

moveCoalesce[_, _, _, {1, c_}, _] := {RND, c}
moveCoalesce[_, _, _, _, {2, d_}] := {RND, d}
moveCoalesce[_, {3, a_}, _, _, _] := {RND, a}
moveCoalesce[_, _, {4, b_}, _, _] := {RND, b}
```

The remaining moveCoalesce rule vacates an occupied site that is not faced by any droplet, and leaves vacant an empty site that is not faced by any droplet.

```
moveCoalesce[_, _, _, _, _] := {0, 0}
```

The moveCoalesce rule set is applied repeatedly to the cloud lattice using the Nest function with the following anonymous function:

```
VonNeumann[moveCoalesce, #]&
```

where

```
VonNeumann[func__, lat_] :=
    MapThread[func, Map[RotateRight[lat, #]&,
              {{0, 0}, {1, 0}, {0, -1}, {-1, 0}, {0, 1}}], 2]
```

These code fragments can be assembled into a program.

The Program

```
stormyWeather[n_, p_, t_]:=
 Module[{RND, cloud, moveCoalesce},

  RND := Random[Integer, {1, 4}];

  cloud = Table[{ RND , 1} * Floor[Random[] + p] , {n}, {n}];

  moveCoalesce[_, {3, a_}, {4, b_}, {1, c_}, {2, d_}] :=
                                      {RND, a + b + c + d};

  moveCoalesce[_, {3, a_}, _, {1, c_}, {2, d_}] := {RND, a + c + d};
  moveCoalesce[_, _, {4, b_}, {1, c_}, {2, d_}] := {RND, b + c + d};
  moveCoalesce[_, {3, a_}, {4, b_}, {1, c_}, _] := {RND, a + b + c};
  moveCoalesce[_, {3, a_}, {4, b_}, _, {2, d_}] := {RND, a + b + d};
```

```
moveCoalesce[_, _, _, {1, c_}, {2, d_}] := {RND, c + d};
moveCoalesce[_, {3, a_}, _, {1, c_}, _] := {RND, a + c};
moveCoalesce[_, _, {4, b_}, {1, c_}, _] := {RND, b + c};
moveCoalesce[_, {3, a_}, _, _, {2, d_}] := {RND, a + d};
moveCoalesce[_, _, {4, b_}, _, {2, d_}] := {RND, b + d};
moveCoalesce[_, {3, a_}, {4, b_}, _, _] := {RND, a + b};

moveCoalesce[_, _, _, {1, c_}, _] := {RND, c};
moveCoalesce[_, _, _, _, {2, d_}] := {RND, d};
moveCoalesce[_, {3, a_}, _, _, _] := {RND, a};
moveCoalesce[_, _, {4, b_}, _, _] := {RND, b};

moveCoalesce[_, _, _, _, _] := {0, 0};

VonNeumann[func__, lat_] :=
    MapThread[func, Map[RotateRight[lat, #]&,
              {{0, 0}, {1, 0}, {0, -1}, {-1, 0}, {0, 1}}], 2];

Nest[VonNeumann[moveCoalesce, #]&, cloud, t]
    ]
```

The Droplet Size Distribution

The size distribution of the droplet in the cloud is easily determined.

We run the stormyWeather program and extract from the resulting matrix of ordered pairs a list of the second components of the ordered pairs.

```
dropletSizeLis =
    Flatten[Partition[Rest[Flatten[stormyWeather [n, p, t]]], 1, 2]]
```

This is a list of the droplet sizes on the lattice. The frequency with which various sized particles occur in dropletSizeLis is given by

```
Function[y, Map[{#, Count[y, #]}&, Range[Max[y]]]][dropletSizeLis]
```

The process of merging droplets is irreversible in the stormyWeather CA. As a result, the average size of the droplets will increase monotonically as the CA evolves until only one droplet (whose size equals the number of droplets in the original cloud) remains in the system.

To obtain a steady-state distribution of coalescing droplet sizes, various mechanisms can be used (Meakin, 1991), including adding (small)

droplets to the system, partially or wholly removing (large) droplets from the system, and fragmenting droplets.

Achieving Steady-State Coalescence

The mechanisms of droplet injection and removal for obtaining a steady-state distribution of droplet sizes (Provata and Nicolis, 1994) can be incorporated into the stormyWeather CA in a direct fashion by modifying (and enlarging) the moveCoalesce rule set, placing various restrictive conditions for pattern-matching the rule arguments. An easier way to implement these mechanisms is to execute each time step in two consecutive half-steps (Takayasu, et al., 1988, Provata and Nicolis, 1994), the first of which moves and coalesces droplets using the moveCoalesce rules, and the second of which injects or removes droplets using the "inject" or "precipitate" rules given below.

Injection

Small droplets (of size 1) can be added to the system at a constant rate by randomly placing the droplets on the sites of the stormyWeather CA lattice at the end of each time step. This is done by replacing the anonymous function in the Nest operation in the stormyWeather program with the anonymous function

```
Map[inject, VonNeumann[moveCoalesce, #], {2}]&
```

where

```
inject[{0, 0}] := {RND, 1} * Floor[Random[] + r]
inject[{x_, y_}] := {x, y + Floor[Random[] + r]}
```

and r is the probability of a droplet being added to a lattice site.

Precipitation

Droplets can be removed from the sites of the stormyWeather lattice when they become sufficiently large (greater than some "criticalSize" value) by replacing the anonymous function in the Nest operation in the stormyWeather program with the anonymous function

```
Map[precipitate, VonNeumann[moveCoalesce, #], {2}]&
```

where

```
precipitate[{x_, y_ /; y > criticalSize}] = {0, 0}
precipitate[z_] = z
```

Computer Simulation Projects

1. Calculate the steady-state droplet size distribution when injection is incorporated into the stormyWeather CA. Calculate the steady-state droplet size distribution when precipitation is incorporated into the model. Modify the CA program to include both injection and precipitation, and determine the steady-state droplet size distribution. Compare the critical exponents for these systems, which have a "source," a "sink," and both a source and a sink, respectively.

2. Modify the precipitate rules to allow for oversized droplets to be partially (rather than wholly) removed from the system at the end of each time step.
   Hint: let the amount of material removed from an oversized droplet be random.

3. The fragmentation of an oversized particle in a coalescence system can be achieved by moving a portion of the particle to the adjacent site it is facing and leaving behind the remainder of the particle facing a randomly chosen direction (Provata and Nicolis, 1994). Implement this mechanism into the stormyWeather program.

References

Bonabeau, Eric and Dogorn, Laurent. 1995. "Possible universality in the size distribution of fish schools." Physical Review E, 51, R5220–R5223.

Meakin, Paul. 1991. "Steady state droplet coalescence." Physica A, 171, 1–18.

Provota, A. and Nicolis, C. 1994. "A microscopic aggregation model of droplet dynamics in warm clouds." J. Statisitical Physics, 74, 75–89.

Takayasu, Hideki, Nishikawa, Ikuko, and Tasaki, Hal. 1988. "Power-law distribution of aggregation systems with injection." Physical Review A, 37, 3110–3117.

11 | Adsorption-Desorption

Introduction

Adsorption-desorption phenomena serve as models of dynamic systems with interesting nonequilibrium phase transition behavior. One example of this kind of system, which is of considerable practical relevance, is the poisoning of a surface during heterogeneous catalysis.

Here, we present a simple CA model of adsorption-desorption, consisting of atoms of a single species that are adsorbed onto vacant surface sites and are desorbed from the surface unless they are completely surrounded by other atoms. We also look at a variant of the model, in which the strength with which an adsorbed atom is bound to the surface is proportional to the number of adsorbed atoms in its vicinity. Finally, we incorporate surface diffusion into the adsorption-desorpton model.

The A Model CA

The A model (Dickman and Burschka, 1988) consists of a single species that adsorbs with some probability on vacant sites and that desorbs with some probability from occupied sites, provided that at least one nearest neighbor site is vacant. When the adsorption probability is small, surface coverage is limited, but when the adsorption probability is large, there is a "poisoning" transition in which the surface becomes completely covered with particles and the system is said to be in the poisoned phase or the adsorbing state.

The System

The system consists of a $(2n + 1)$ by $(2n + 1)$ square lattice with periodic boundary conditions. The lattice sites have value 0, indicating an empty site, or 1, indicating an occupied site. Initially, all of the sites are occupied except for the center site, which is vacant.

```
initConf =
    ReplacePart[Table[1, {2n + 1}, {2n + 1}], 0, {n + 1, n + 1}]
```

The Update Rules

The update rules for this system have five arguments: the value of a site and the values of the nearest neighbor sites lying north, east, south, and west of the site.

- An occupied site remains occupied (i.e., an adsorbed particle remains adsorbed) if all of its nearest neighbor sites are occupied.

```
ADR[1, 1, 1, 1, 1] := 1
```

- An empty site or an occupied site with at least one empty nearest neighbor site becomes or remains empty with probability $(1 - p)$, and becomes or remains occupied with probability p.

```
ADR[_, _, _, _, _] := Floor[Random[] + p]
```

Note: This rule could be written as a set of 32 rules using specific argument values of 1's and 0's rather than with underspaces, but this would make the program less readable without increasing its efficiency.

Applying the Rules

The update rules are applied to the CA lattice using the anonymous function

```
VonNeumann[ADR, #]&
```

where

```
VonNeumann[func__, lat_] :=
    MapThread[func, Map[RotateRight[lat, #]&,
            {{0, 0}, {1, 0}, {0, -1}, {-1, 0}, {0, 1}}], 2]
```

The lattice is updated over t time steps by applying this anonymous function using the Nest function

```
Nest[VonNeumann[ADR, #]&, initConf, t]
```

The Program

```
Amodel[p_, n_, t_] :=
 Module[{initConf, ADR, VonNeumann},

  initConf =
        ReplacePart[Table[1, {2n + 1}, {2n + 1}], 0, {n + 1, n + 1}];

  ADR[1, 1, 1, 1, 1] := 1;
  ADR[_, _, _, _, _] := Floor[Random[] + p];

  VonNeumann[func__, lat_] :=
        MapThread[func, Map[RotateRight[lat, #]&,
                 {{0, 0}, {1, 0}, {0, -1}, {-1, 0}, {0, 1}}], 2];

  Nest[VonNeumann[ADR, #]&, initConf, t]
 ]
```

The CPM Model CA

The contact process model (Beney, et al., 1990) differs from the A model in that the probability of desorption decreases as the number of nearest neighbors increases.

The Update Rules

- An empty site becomes occupied by a particle (i.e., a particle is adsorbed) with probability p and remains empty with a probability $(1 - p)$.

```
CPM[0, _, _, _, _] := Floor[Random[] + p]
```

- An occupied site becomes vacant (i.e., a particle is desorbed) with probability $q*$ (fraction of vacant nearest neighbor sites) and remains occupied with a probability $1 - q*$ (fraction of vacant nearest neighbor sites).

```
CPM[1, 1, 1, 1, 1] := 1

CPM[1, a_, b_, c_, d_] :=
                    Floor[Random[] + 1 - q * (4 - (a + b + c + d))/4]
```

Note: While the CPM[1, a_, b_, c_, d_] rule could be used to calculate the case of CPM[1, 1 ,1, 1, 1], a significant speed increase is obtained by writing a rule specifically for that case.

The Program

```
ContactProcessModel[n_, p_, q_, t_]:=
 Module[{initConf, CPM, VonNeumann},

  initConf =
        ReplacePart[Table[1, {2n + 1}, {2n + 1}], 0, {n + 1, n + 1}];

  CPM[0, _, _, _, _] := Floor[Random[] + p];
  CPM[1, a_, b_, c_, d_] :=
                Floor[Random[] + 1 - q * (4 - (a + b + c + d))/4];
  CPM[1, 1, 1, 1, 1] := 1;

  VonNeumann[func__, lat_] :=
      MapThread[func, Map[RotateRight[lat, #]&,
                {{0, 0}, {1, 0}, {0, -1}, {-1, 0}, {0, 1}}], 2];

  Nest[VonNeumann[CPM, #]&, initConf, t]
 ]
```

Adsorption-Desorption with Surface Diffusion

When atoms are adsorbed on a surface, they need not be frozen in place until they desorb. Instead, adsorbed atoms can undergo random (Brownian) motion along the surface. This movement is expected to raise the critical value of the adsorption probability at which surface coverage becomes complete.

A simple way to incorporate the diffusion of atoms along the surface into the A model CA (Bagnoli, et al., 1992) is to first apply the adsorption-desorption rules one time and then apply rules for the diffusion of particles on the lattice r times, on each time step (the Solidification CA takes the same approach for the diffusion of heat). To do this, we can employ the CA rule set for multiple random walkers with excluded volume.

Note: when r equals zero, the A model is recovered.

To combine the random walkers and A model CAs, we need to replace the integer 1 in the A model with the integers 1 through 4 where these values represent the following quantities:

1 — a site occupied by a north-facing walker
2 — a site occupied by an east-facing walker
3 — a site occupied by a south-facing walker
4 — a site occupied by a west-facing walker

The integer 0 retains its meaning as an empty site.

The System

Initially, the sites of an *n*-by-*n* square lattice with periodic boundary conditions are randomly occupied with probability *p* by random walkers facing randomly chosen directions.

```
initConf =
        Table[Random[Integer, {1, 4}] * Floor[p + Random[]], {n}, {n}]
```

Adsorption-Desorption

The rules for adsorption-desorption have the same form as the two rules used in the A model:

```
ADR[_, _, _, _, _] := Random[Integer, {1, 4}] * Floor[Random[] + p]

ADR[_?Positive, _?Positive, _?Positive, _?Positive, _?Positive] :=
                                    Random[Integer, {1, 4}]
```

Surface Diffusion

The rules for surface diffusion take 13 arguments, representing the value of the site, the values of the four nearest neighbors in the N, E, S, W directions, the values of the four nearest neighbors in the NE, SE, SW, NW directions, and the values of four next nearest neighbors in the N, E, S, W directions.

The following 28 rules are used:

```
RND := Random[Integer, {1, 4}]
diffuse[1, 0, _, _, _, 4, _, _, _, _, _, _, _] := RND
diffuse[1, 0, _, _, _, _, _, _, 2, _, _, _, _] := RND
diffuse[1, 0, _, _, _, _, _, _, _, 3, _, _, _] := RND
diffuse[1, 0, _, _, _, _, _, _, _, _, _, _, _] := 0
diffuse[2, _, 0, _, _, 3, _, _, _, _, _, _, _] := RND
diffuse[2, _, 0, _, _, _, 1, _, _, _, _, _, _] := RND
diffuse[2, _, 0, _, _, _, _, _, _, 4, _, _, _] := RND
diffuse[2, _, 0, _, _, _, _, _, _, _, _, _, _] := 0
diffuse[3, _, _, 0, _, _, 4, _, _, _, _, _, _] := RND
diffuse[3, _, _, 0, _, _, _, 2, _, _, _, _, _] := RND
diffuse[3, _, _, 0, _, _, _, _, _, _, _, 1, _] := RND
diffuse[3, _, _, 0, _, _, _, _, _, _, _, _, _] := 0
diffuse[4, _, _, _, 0, _, _, 1, _, _, _, _, _] := RND
diffuse[4, _, _, _, 0, _, _, _, 3, _, _, _, _] := RND
diffuse[4, _, _, _, 0, _, _, _, _, _, _, _, 2] := RND
diffuse[4, _, _, _, 0, _, _, _, _, _, _, _, _] := 0
```

```
diffuse[_?Positive, _, _, _, _, _, _, _, _, _, _, _, _] := RND
diffuse[0, 3, 4, _, _, _, _, _, _, _, _, _, _] := 0
diffuse[0, 3, _, 1, _, _, _, _, _, _, _, _, _] := 0
diffuse[0, 3, _, _, 2, _, _, _, _, _, _, _, _] := 0
diffuse[0, _, 4, 1, _, _, _, _, _, _, _, _, _] := 0
diffuse[0, _, 4, _, 2, _, _, _, _, _, _, _, _] := 0
diffuse[0, _, _, 1, 2, _, _, _, _, _, _, _, _] := 0
diffuse[0, 3, _, _, _, _, _, _, _, _, _, _, _] := RND
diffuse[0, _, 4, _, _, _, _, _, _, _, _, _, _] := RND
diffuse[0, _, _, 1, _, _, _, _, _, _, _, _, _] := RND
diffuse[0, _, _, _, 2, _, _, _, _, _, _, _, _] := RND
diffuse[0, _, _, _, _, _, _, _, _, _, _, _, _] := 0
```

Applying the Update Rules

In a given time step, we first apply the adsorption-desorption rules to the CA lattice, which we'll call mat, one time, using

```
VonNeumann[ADR, mat]
```

where

```
VonNeumann[func__, lat_] :=
    MapThread[func, Map[RotateRight[lat, #]&,
            {{0, 0}, {1, 0}, {0, -1}, {-1, 0}, {0, 1}}], 2]
```

and then we apply the diffusion rules r times using

```
Nest[MvonN[diffuse, #]&, VonNeumann[ADR, mat], r]
```

where

```
MvonN[func__, lat_] :=
   MapThread[func, Map[RotateRight[lat, #]&,
            {{0, 0}, {1, 0}, {0, -1}, {-1, 0}, {0, 1},
             {1, -1}, {-1, -1}, {-1, 1}, {1, 1},
             {2, 0}, {0, -2}, {-2, 0}, {0, 2}}], 2]
```

The sites in the lattice are updated over t time steps, starting with the initial lattice configuration, using the following operation:

```
NestList[Nest[MvonN[diffuse, #]&, VonNeumann[ADR, #], r]&,
        initConf, t]
```

The Program

```
ADwithDiffusion[p_, n_, r_, t_] :=
Module[{initConf, ADR, RND, diffuse, VonNeumann, MvonN},

 initConf =
     Table[Random[Integer, {1, 4}] * Floor[p + Random[]], {n}, {n}];


 ADR[_, _, _, _, _] := Random[Integer, {1, 4}] * Floor[Random[] + p];
 ADR[_?Positive, _?Positive, _?Positive, _?Positive, _?Positive] :=
                                     Random[Integer, {1, 4}];
 RND := Random[Integer, {1, 4}];


 diffuse[1, 0, _, _, _, 4, _, _, _, _, _, _, _] := RND;
 diffuse[1, 0, _, _, _, _, _, _, 2, _, _, _, _] := RND;
 diffuse[1, 0, _, _, _, _, _, _, _, 3, _, _, _] := RND;
 diffuse[1, 0, _, _, _, _, _, _, _, _, _, _, _] := 0;
 diffuse[2, _, 0, _, _, 3, _, _, _, _, _, _, _] := RND;
 diffuse[2, _, 0, _, _, _, 1, _, _, _, _, _, _] := RND;
 diffuse[2, _, 0, _, _, _, _, _, _, 4, _, _] := RND;
 diffuse[2, _, 0, _, _, _, _, _, _, _, _, _] := 0;
 diffuse[3, _, _, 0, _, _, 4, _, _, _, _, _] := RND;
 diffuse[3, _, _, 0, _, _, _, 2, _, _, _, _] := RND;
 diffuse[3, _, _, 0, _, _, _, _, _, _, 1, _] := RND;
 diffuse[3, _, _, 0, _, _, _, _, _, _, _, _] := 0;
 diffuse[4, _, _, _, 0, _, _, 1, _, _, _, _] := RND;
 diffuse[4, _, _, _, 0, _, _, _, 3, _, _, _] := RND;
 diffuse[4, _, _, _, 0, _, _, _, _, _, _, 2] := RND;
 diffuse[4, _, _, _, 0, _, _, _, _, _, _, _] := 0;
 diffuse[_?Positive, _, _, _, _, _, _, _, _, _, _, _] := RND;
 diffuse[0, 3, 4, _, _, _, _, _, _, _, _, _] := 0;
 diffuse[0, 3, _, 1, _, _, _, _, _, _, _, _] := 0;
 diffuse[0, 3, _, _, 2, _, _, _, _, _, _, _] := 0;
 diffuse[0, _, 4, 1, _, _, _, _, _, _, _, _] := 0;
 diffuse[0, _, 4, _, 2, _, _, _, _, _, _, _] := 0;
 diffuse[0, _, _, 1, 2, _, _, _, _, _, _, _] := 0;
 diffuse[0, 3, _, _, _, _, _, _, _, _, _, _] := RND;
 diffuse[0, _, 4, _, _, _, _, _, _, _, _, _] := RND;
 diffuse[0, _, _, 1, _, _, _, _, _, _, _, _] := RND;
 diffuse[0, _, _, _, 2, _, _, _, _, _, _, _] := RND;
 diffuse[0, _, _, _, _, _, _, _, _, _, _, _] := 0;


 VonNeumann[func__, lat_] :=
     MapThread[func, Map[RotateRight[lat, #]&,
             {{0, 0}, {1, 0}, {0, -1}, {-1, 0}, {0, 1}}], 2];
```

```
MvonN[func__, lat_] :=
  MapThread[func, Map[RotateRight[lat, #]&,
            {{0, 0}, {1, 0}, {0, -1}, {-1, 0}, {0, 1},
             {1, -1}, {-1, -1}, {-1, 1}, {1, 1},
             {2, 0}, {0, -2}, {-2, 0}, {0, 2}}], 2];


  NestList[Nest[MvonN[diffuse, #]&, VonNeumann[ADR, #], r]&,
           initConf, t]
]
```

Computer Simulation Project

As the adsorption probability increases, a second-order transition occurs at p_c such that the fraction of covered (i.e., occupied) sites, $x_A < 1$ for $p < p_c$ and $x_A = 1$ for $p >= p$. Determine the value of p_c and the value of the critical exponent, b, in the relationship

$$(1 - x_A) = (p_c - p)^b$$

for each of the three models of adsorption-desorption presented above.

References

Bagnoli, Franco, Chopard, Bastien, Droz, Michael and Frachebourg, Laurent. 1992. "Critical behavior of a diffusive model with one adsorbing state." J. Phys. A, 25, 1085–1091.

Beney, Phillipe, Droz, Michael and Frachebourg, Laurent. 1990. "On the critical behavior of cellular automata models of non-equilibrium phase transitions." J. Phys. A, 23, 3353–3359.

Chopard, Bastien, Phillipe, Droz, Michael, Cornell, Stephen and Frachebourg, Laurent. "Cellular automata approach to reaction-diffusion systems: theory and application." In *Cellular Automata Prospects in Astrophysical Applications*, Perdang, J. M. and Lejeune, editors, World Scientific (1995), 157–186.

Dickman, Ronald and Burschka, Martin A. 1988. "Nonequilibrium critical poisoning in a single-species model." Physics Letters A, 127, 132–137.

12 Chemotaxis

Introduction

Chemical signaling plays a major role in communications among the members of animal, bird, insect, and even bacterial societies (it has also been speculated that some sort of chemical signaling occurs between humans) and these signals influence various sorts of social behavior (e.g., mating and aggression) in these systems (Pennisi, 1995). Movement, in particular, is often governed by chemicals. For example, the production and emission of cAMP by slime mold amoeba and the attraction of the amoeba to this chemical brings about their aggregation into clusters when there is a scarcity of food, while the mass recruitment of ants to the collective task of food foraging employs the emission of pheromones by those ants carrying food to lead other ants to the food source.

We will present a simple random walkers CA model (Resnick, 1994) of chemotaxis, the movement in response to a chemical gradient, for the case where the chemical that attracts the walkers is emitted by each walker as it moves, and is dissipated into the environment in the absense of a walker.

The Chemotaxis CA

The System

The system consists of an n-by-n square lattice with periodic boundary conditions. Each lattice site has a value that is an ordered pair. The first component of the ordered pair is a nonnegative integer that represents the level of chemical attractant on the site. The second component of the ordered pair is an integer between 0 and 4 where the values have the following meanings:

0 — an empty site
1 — a north-facing ant
2 — an east-facing ant
3 — a south-facing ant
4 — a west-facing ant

Initially, all of the sites have a zero level of chemical attractant and the sites are randomly occupied with probability p by random walkers facing randomly chosen directions.

```
initConf =
  Table[{0, Random[Integer, {1, 4}]} * Floor[Random[] + p], {n}, {n}]
```

The Update Rules

A simple way to combine the effect of an "active landscape" on the movement of the ants is to first apply the rules for orienting the ant and then apply the rules for diffusing the ant on the lattice, on each time step.

Orienting the Walkers

The rules for orienting an ant take five arguments:

```
sniff[site, N, E, S, W]
```

where the five arguments represent the value of the site and the values of the four nearest neighbors in the North, East, South, and West directions, respectively.
There are two rules:

- An empty site remains unchanged.

```
sniff[{a_, 0}, _, _, _, _] := {a, 0}
```

- An ant orients itself so as to face the nearest neighbor site that has the highest chemical attractant level. If two or more nearest neighbor sites have the highest level of chemical attractant, the ant randomly chooses one of these sites to face.

```
sniff[{a_, _}, {n_, _}, {e_, _}, {s_, _}, {w_, _}] :=
  {a, #[[Random[Integer, {1, Length[#]}]]]}&[Flatten[Position[#,
                                              Max[#]]]&[{n, e, s, w}]]}
```

Taking a Step

The rules for ant movement take 13 arguments:

```
walk[site, N, E, S, W, NE, SE, SW, NW, Nn, Ee, Ss, Ww]
```

where the arguments represent the value of the site, the values of the four nearest neighbors in the N, E, S, W directions, the values of the four nearest neighbors in the NE, SE, SW, NW directions, and the values of four next nearest neighbors in the N, E, S, W directions.

When an ant moves, it leaves behind a chemical scent to attract other ants. In the absence of an ant, the chemical scent that has been laid down dissipates.

These changes in the strength of the chemical scent are easily expressed in rewrite rules by using the 28 update rules of the multiple walkers CA with minor modifications.

In the following rules, the chemical scent on an occupied site that is vacated by an ant is increased by 1, while the chemical scent on an unoccupied site that remains empty is reduced by 1.

```
walk[{a_, 1}, {_, 0}, _, _, _, {_, 4}, _, _, _, _, _, _, _] := {a, 1}
walk[{a_, 1}, {_, 0}, _, _, _, _, _, _, {_, 2}, _, _, _, _] := {a, 1}
walk[{a_, 1}, {_, 0}, _, _, _, _, _, _, _, {_, 3}, _, _, _] := {a, 1}
walk[{a_, 1}, {_, 0}, _, _, _, _, _, _, _, _, _, _, _] := {a + 1, 0}
walk[{a_, 2}, _, {_, 0}, _, _, {_, 3}, _, _, _, _, _, _, _] := {a, 2}
walk[{a_, 2}, _, {_, 0}, _, _, _, {_, 1}, _, _, _, _, _, _] := {a, 2}
walk[{a_, 2}, _, {_, 0}, _, _, _, _, _, _, {_, 4}, _, _] := {a, 2}
walk[{a_, 2}, _, {_, 0}, _, _, _, _, _, _, _, _, _, _] := {a + 1, 0}
walk[{a_, 3}, _, _, {_, 0}, _, _, {_, 4}, _, _, _, _, _, _] := {a, 3}
walk[{a_, 3}, _, _, {_, 0}, _, _, _, {_, 2}, _, _, _, _, _] := {a, 3}
walk[{a_, 3}, _, _, {_, 0}, _, _, _, _, _, _, _, {_, 1}, _] := {a, 3}
walk[{a_, 3}, _, _, {_, 0}, _, _, _, _, _, _, _, _, _] := {a + 1, 0}
walk[{a_, 4}, _, _, _, {_, 0}, _, _, {_, 1}, _, _, _, _, _] := {a, 4}
walk[{a_, 4}, _, _, _, {_, 0}, _, _, _, {_, 3}, _, _, _, _] := {a, 4}
walk[{a_, 4}, _, _, _, {_, 0}, _, _, _, _, _, _, _, {_, 2}] := {a, 4}
walk[{a_, 4}, _, _, _, {_, 0}, _, _, _, _, _, _, _, _] := {a + 1, 0}
walk[{a_, b_}, _, _, _, _, _, _, _, _, _, _, _, _] := {a + 1, b}
walk[{a_, 0}, {_, 3}, {_, 4}, _, _, _, _, _, _, _, _, _, _] :=
                                                    {Max[a - 1, 0], 0}
walk[{a_, 0}, {_, 3}, _, {_, 1}, _, _, _, _, _, _, _, _, _] :=
                                                    {Max[a - 1, 0], 0}
walk[{a_, 0}, {_, 3}, _, _, {_, 2}, _, _, _, _, _, _, _, _] :=
                                                    {Max[a - 1, 0], 0}
```

```
walk[{a_, 0}, _, {_, 4}, {_, 1}, _, _, _, _, _, _, _, _, _] :=
                                                        {Max[a - 1, 0], 0}
walk[{a_, 0}, _, {_, 4}, _, {_, 2}, _, _, _, _, _, _, _, _] :=
                                                        {Max[a - 1, 0], 0}
walk[{a_, 0}, _, _, {_, 1}, {_, 2}, _, _, _, _, _, _, _, _] :=
                                                        {Max[a - 1, 0], 0}
walk[{a_, 0}, {_, 3}, _, _, _, _, _, _, _, _, _, _, _] := {a, 3}
walk[{a_, 0}, _, {_, 4}, _, _, _, _, _, _, _, _, _, _] := {a, 4}
walk[{a_, 0}, _, _, {_, 1}, _, _, _, _, _, _, _, _, _] := {a, 1}
walk[{a_, 0}, _, _, _, {_, 2}, _, _, _, _, _, _, _, _] := {a, 2}
walk[{a_, 0}, _, _, _, _, _, _, _, _, _, _, _, _] :=
                                                        {Max[a - 1, 0], 0}
```

Note: We have arbitrarily chosen to assume that an ant retains its orientation when it attempts to move. However, the specific non-zero value that is used for the second component of the ordered pair is irrelevant since, as we will show in the next section, each ant chooses its orientation (using the sniff rules) immediately after attempting a move.

Applying the Update Rules

In a given time step, the sniff rules are first applied to the CA lattice, which we'll call mat, using

```
VonNeumann[sniff, #]&[mat]
```

where

```
VonNeumann[func__, lat_] :=
    MapThread[func, Map[RotateRight[lat, #]&,
              {{0, 0}, {1, 0}, {0, -1}, {-1, 0}, {0, 1}}]], 2]
```

The diffusion rules are then applied once, using

```
MvonN[walk, #]&[VonNeumann[sniff, #]&[mat]]
```

where

```
MvonN[func__, lat_] :=
  MapThread[func, Map[RotateRight[lat, #]&,
            {{0, 0}, {1, 0}, {0, -1}, {-1, 0}, {0, 1},
             {1, -1}, {-1, -1}, {-1, 1}, {1, 1},
             {2, 0}, {0, -2}, {-2, 0}, {0, 2}}]], 2]
```

The sites in the lattice are updated over *t* time steps, starting with the initial lattice configuration, using the following operation:

```
NestList[MvonN[walk, #]&[VonNeumann[sniff, #]]&, initConf, t]
```

Note: It is also permissible to apply the walk rules before the sniff rules.

The Program

```
chemotaxis[m_, p_, t_] :=
Module[{initConf, sniff, RND, walk, VonNeumann, MvonN},

 initConf =
  Table[{0, Random[Integer, {1, 4}]} * Floor[Random[] + p], {n}, {n}];

 sniff[{a_, 0}, _, _, _, _] := {a, 0};
 sniff[{a_, _}, {n_, _}, {e_, _}, {s_, _}, {w_, _}] :=
  {a, #[[Random[Integer, {1, Length[#]}]]]}&[Flatten[Position[#,
                                    Max[#]]]&[{n, e, s, w}]]};

 walk[{a_, 1}, {_, 0}, _, _, _, {_, 4}, _, _, _, _, _, _, _] := {a, 1};
 walk[{a_, 1}, {_, 0}, _, _, _, _, _, _, {_, 2}, _, _, _, _] := {a, 1};
 walk[{a_, 1}, {_, 0}, _, _, _, _, _, _, _, {_, 3}, _, _, _] := {a, 1};
 walk[{a_, 1}, {_, 0}, _, _, _, _, _, _, _, _, _, _, _] := {a + 1, 0};
 walk[{a_, 2}, _, {_, 0}, _, _, {_, 3}, _, _, _, _, _, _, _] := {a, 2};
 walk[{a_, 2}, _, {_, 0}, _, _, _, {_, 1}, _, _, _, _, _, _] := {a, 2};
 walk[{a_, 2}, _, {_, 0}, _, _, _, _, _, _, _, {_, 4}, _, _] := {a, 2};
 walk[{a_, 2}, _, {_, 0}, _, _, _, _, _, _, _, _, _, _] := {a + 1, 0};
 walk[{a_, 3}, _, _, {_, 0}, _, _, {_, 4}, _, _, _, _, _] := {a, 3};
 walk[{a_, 3}, _, _, {_, 0}, _, _, _, {_, 2}, _, _, _, _] := {a, 3};
 walk[{a_, 3}, _, _, {_, 0}, _, _, _, _, _, _, {_, 1}, _] := {a, 3};
 walk[{a_, 3}, _, _, {_, 0}, _, _, _, _, _, _, _, _] := {a + 1, 0};
 walk[{a_, 4}, _, _, _, {_, 0}, _, _, {_, 1}, _, _, _, _] := {a, 4};
 walk[{a_, 4}, _, _, _, {_, 0}, _, _, _, {_, 3}, _, _, _] := {a, 4};
 walk[{a_, 4}, _, _, _, {_, 0}, _, _, _, _, _, _, {_, 2}] := {a, 4};
 walk[{a_, 4}, _, _, _, {_, 0}, _, _, _, _, _, _, _] := {a + 1, 0};
 walk[{a_, b_}, _, _, _, _, _, _, _, _, _, _, _, _] := {a + 1, b};
 walk[{a_, 0}, {_, 3}, {_, 4}, _, _, _, _, _, _, _, _, _] :=
                                    {Max[a - 1, 0], 0};
 walk[{a_, 0}, {_, 3}, _, {_, 1}, _, _, _, _, _, _, _, _] :=
                                    {Max[a - 1, 0], 0};
 walk[{a_, 0}, {_, 3}, _, _, {_, 2}, _, _, _, _, _, _, _] :=
                                    {Max[a - 1, 0], 0};
```

```
walk[{a_, 0}, _, {_, 4}, {_, 1}, _, _, _, _, _, _, _, _, _] :=
                                              {Max[a - 1, 0], 0};
walk[{a_, 0}, _, {_, 4}, _, {_, 2}, _, _, _, _, _, _, _, _] :=
                                              {Max[a - 1, 0], 0};
walk[{a_, 0}, _, _, {_, 1}, {_, 2}, _, _, _, _, _, _, _, _] :=
                                              {Max[a - 1, 0], 0};
walk[{a_, 0}, {_, 3}, _, _, _, _, _, _, _, _, _, _, _] := {a, 3};
walk[{a_, 0}, _, {_, 4}, _, _, _, _, _, _, _, _, _, _] := {a, 4};
walk[{a_, 0}, _, _, {_, 1}, _, _, _, _, _, _, _, _, _] := {a, 1};
walk[{a_, 0}, _, _, _, {_, 2}, _, _, _, _, _, _, _, _] := {a, 2};
walk[{a_, 0}, _, _, _, _, _, _, _, _, _, _, _, _] :=
                                              {Max[a - 1, 0], 0};


VonNeumann[func__, lat_] :=
    MapThread[func, Map[RotateRight[lat, #]&,
              {{0, 0}, {1, 0}, {0, -1}, {-1, 0}, {0, 1}}], 2];


MvonN[func__, lat_] :=
  MapThread[func, Map[RotateRight[lat, #]&,
            {{0, 0}, {1, 0}, {0, -1}, {-1, 0}, {0, 1},
             {1, -1}, {-1, -1}, {-1, 1}, {1, 1},
             {2, 0}, {0, -2}, {-2, 0}, {0, 2}}], 2];


NestList[MvonN[walk, #]&[VonNeumann[sniff, #]]&, initConf, t]
]
```

Following the Trail

As the ants move, they will tend to follow trails laid down by other ants. This can result in the clustering together of the ants. We can look at the paths taken by the ants using a `RasterArray` graphics whose elements are colored according to the amount of chemical attractant on the CA lattice site. Below are shown the results obtained for various densities of ants on a 50-by-50 lattice over 300 time steps.

```
SeedRandom[7]
(attractantMap = chemotaxis[50, 0.05, 300] /. {x_Integer, _} -> x);

Show[GraphicsArray[
  Partition[Map[Show[Graphics[
    RasterArray[# /. Thread[Range[0, Max[#]] ->
    Map[Hue, Table[Random[], {Max[#] + 1}]]]]],
   DisplayFunction -> Identity, AspectRatio -> Automatic] ]&,
    attractantMap[[Range[1, 301, 100 ] ]]], 2] ]];
```

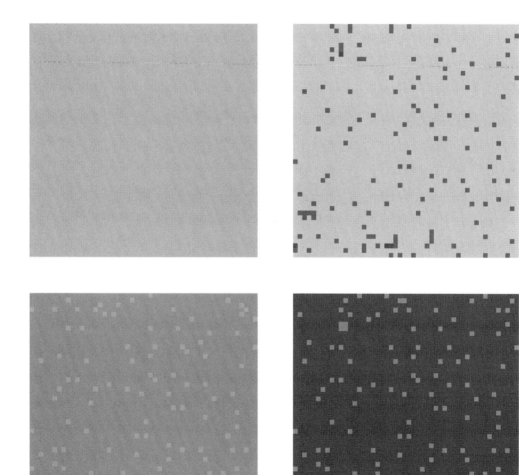

```
SeedRandom[7]
(attractantMap = chemotaxis[50, 0.10, 300] /. {x_Integer, _} -> x);

Show[GraphicsArray[
  Partition[Map[Show[Graphics[
    RasterArray[# /. Thread[Range[0, Max[#]] ->
    Map[Hue, Table[Random[], {Max[#] + 1}]]]]],
    DisplayFunction -> Identity, AspectRatio -> Automatic] ]&,
    attractantMap[[Range[1, 301, 100 ] ]]], 2] ]];
```

```
SeedRandom[7]
(attractantMap = chemotaxis[50, 0.25, 300] /. {x_Integer, _} -> x);

Show[GraphicsArray[
  Partition[Map[Show[Graphics[
    RasterArray[# /.
        Thread[Range[0, Max[#]] ->
            Map[Hue, Table[Random[], {Max[#] + 1}]]]]],
                    DisplayFunction -> Identity,
                    AspectRatio -> Automatic]]&,
            attractantMap[[Range[1, 301, 100]]]], 2]]];
```

Computer Simulation Project

Modify the sniff rules so that an ant faces the empty nearest neighbor site with the highest level of chemical attractant, and when there are no empty nearest neighbor sites, the ant retains its current orientation.

References

Freund, Harald and Grassberger, Peter. 1992. "The red queen's walk." Physica A, 190, 218–237.

Lam, Lui and Pochy, Rocco. 1993. "Active-walker models: growth and form in nonequilibrium systems." Computers in Physics, 7, 534–541.

Pennisi, Elizabeth. 1995. "The secret language of bacteria.' The New Scientist, September 16, 30–33.

Resnick, Mitchel. *Turtles, Termites, and Traffic Jams*. MIT Press (1994), 50–68.

13 Ant Colony Activity

Introduction

While an individual ant is considered by many people to have no significance (hence the phrase "crush like an ant"), the complex behavior of a "society" of ants is rather fascinating (the so-called ant farm has long been popular among scientifically inclined young people). Various studies (Goodwin, 1994, Gordon, 1995) have shown that an individual ant cannot be trained to perform even the simplest of tasks and that groups of ants have no hierarchical order or chain of command by which one ant directs the action of another ant. In the absence of native intelligence and social control, the question is how does the collective behavior carried out in a colony of ants emerge from the activity of the individual ants in the colony? It has been suggested (Goodwin, 1994, Gordon, 1995) that local interactions between ants, of the sort found in a cellular automaton, suffice to produce the observed organizational features of an ant colony. We will develop a CA that models the activity level of an ant colony over time as a function of the size of the colony.

The Ant Colony CA

The model employs a square lattice with reflecting boundary conditions whose interior sites are either empty or occupied by an ant that is in either an active or inactive state. The following statement describes how the sites in the lattice evolve over time (Goodwin, 1994):

"When in an active state, individuals move from one site to any adjacent, unoccupied site on a lattice that defines the territory of the colony. If an individual moves to a site adjacent to one occupied by an inactive ant, the latter is stimulated to become active, but its activity will cease after a time unless it becomes spontaneously active or is simulated by another ant."

Unfortunately, this description is a bit vague and it is necessary to be more precise in specifying the rules of an ant colony cellular automaton.

In developing our ant colony CA program, we assume (based on what seems physically reasonable) that the lattice sites in the system are updated according to the following rules (it is straightforward to modify our program for an alternative interpretation of the ant colony model):

- An active ant moves to the adjacent empty site it faces, unless the move will result in a collision with another active ant.
- An active ant facing an adjacent site that is either on the border or is occupied by another ant stays put.
- An active ant, whether it moves or not, randomly picks a direction to face in each time step and increments its "activity time clock" (which keeps track of how many time steps have elapsed since the ant became active) by 1.
- An active ant becomes inactive after s time steps.
- An inactive ant becomes active either spontaneously or when an active ant is on an adjacent site.

The System

The system consists of an n-by-n square lattice whose sites have values that are ordered pairs $\{x, y\}$.

The first component of the ordered pair is either the symbol, b, or an integer ranging from -1 to 4 where the values have the following meanings:

b — a border site
0 — an empty site
-1 — a site occupied by an inactive ant
1 — a site occupied by a north-facing active ant
2 — a site occupied by an east-facing active ant
3 — a site occupied by a south-facing active ant
4 — a site occupied by a west-facing active ant

The second component of the ordered pair is an integer ranging from 0 to s where the value is an *activity time clock* indicating the number of time steps that the ant has been in the active state. A site that is empty, on the border, or has an inactive ant on it, has an activity time clock value of 0. A site with an active ant on it has a value of $1, 2, \ldots, s$.

We start by randomly placing inactive and active ants on the sites of an $(n-1)$-by-$(n-1)$ lattice, with a probability p. If the ant is active, it faces a randomly chosen direction and its time clock is randomly set between 1 and s.

```
antPopulation =
  Table[{{-1, 1, 2, 3, 4}[[Random[Integer, {1, 5}] ]],
        Random[Integer, {1, s}]}* Floor[Random[] + p],
            {n - 1}, {n - 1}] /. {-1, _} -> {-1, 0}
```

We decorate the antPopulation lattice with a border of sites having value $\{b, 0\}$ on the left and bottom sides of the lattice using

```
antFarm = border[antPopulation]
```

where

```
border =
  Append[Map[Append[#, {b, 0}]&, #], Table[{b, 0}, {Length[#] + 1}]]&
```

Note: It's not necessary to place borders along all four sides of the lattice (this is analogous to the use of one rather two borders in the inter-facial diffusion front CA). The choice of which column (left or right) and which row (top or bottom) are decorated with a border is irrelevant.

The ant density on the ant farm is conserved in the system over time. It can be calculated using

```
antDensity =
  N[Apply[Plus,
          Abs[Sign[Transpose[Flatten[antPopulation,1]][[1]]]]]]/n^2]
```

When the system is large, the ant density approaches the value of p.

The Rules

The rules for the behavior of an ant during a time step, take 13 arguments:

```
ant[site, N, E, S, W, NE, SE, SW, NW, Nn, Ee, Ss, Ww]
```

The arguments represent the value of the site and the values of 12 neighbor sites: the nearest neighbors in the N, E, S, W directions, the nearest neighbors in the NE, SE, SW, NW directions, and the next nearest neighbors in the N, E, S, W directions (these values are needed to implement the "excluded volume" constraint).

The following rules describe the behavior of sites during a time step:

The function RND is a random number generator used to determine the direction faced by an active ant.

```
RND := Random[Integer, {1, 4}]
```

For an inactive ant (a site whose value is $\{-1, 0\}$):

- An inactive ant adjacent to an active ant becomes active, randomly chooses a direction to face, and resets its activity time clock to 1.

```
ant[{-1,  0}, {x_?Positive, _}, _, _, _, _, _, _, _, _, _, _, _] :=
                                                              {RND, 1}
ant[{-1,  0},  _, {x_?Positive,_}, _, _, _, _, _, _, _, _, _, _] :=
                                                              {RND, 1}
ant[{-1,  0},  _, _, {x_?Positive,_}, _, _, _, _, _, _, _, _, _] :=
                                                              {RND, 1}
ant[{-1,  0},  _, _, _, {x_?Positive,_}, _, _, _, _, _, _, _, _] :=
                                                              {RND, 1}
```

- An inactive ant becomes active with probability r, randomly chooses a direction to face, and resets its activity time clock to 1.

```
ant[{-1, 0}, _, _, _, _, _, _, _, _, _, _, _, _] :=
        { -1 + # * Random[Integer, {2, 5}], #}&[Floor[Random[] + r] ]
```

For an active ant (a site whose value is an ordered pair whose first component is 1, 2, 3 or 4):

- An active ant become inactive after s time steps and resets its activity time clock to 0.

```
ant[{x_?Positive, s}, _, _, _, _, _, _, _, _, _, _, _, _] := {-1, 0}
```

- An active ant facing an adjacent site that is either on the border or is occupied by another ant, stays put, randomly chooses a direction to face, and increments its activity time clock by 1.

```
ant[{1, y_}, {-1 | x_?Positive | b, _}, _, _, _, _,
                              _, _, _, _, _, _, _] := {RND, y + 1}
ant[{2, y_}, _, {-1 | x_?Positive | b, _}, _, _, _,
                              _, _, _, _, _, _, _] := {RND, y + 1}
ant[{3, y_}, _, _, {-1 | x_?Positive | b, _}, _, _,
                              _, _, _, _, _, _, _] := {RND, y + 1}
ant[{4, y_}, _, _, _, {-1 | x_?Positive | b, _}, _,
                              _, _, _, _, _, _, _] := {RND, y + 1}
```

- An active ant facing an empty adjacent site that is also faced by another active ant (e.g., an active right-facing ant that lies to the left of an empty site that is faced by an active left-facing ant lying to the right of the empty site) stays put, randomly chooses a direction to face, and increments its activity time clock by 1.

```
ant[{1, y_}, {0, 0}, _, _, _, {4, _}, _, _, _, _, _, _, _] :=
                                                        {RND, y + 1}
ant[{1, y_}, {0, 0}, _, _, _, _, _, _, {2, _}, _, _, _, _] :=
                                                        {RND, y + 1}
ant[{1, y_}, {0, 0}, _, _,_ ,_ ,_ ,_ , _, {3, _}, _, _, _] :=
                                                        {RND, y + 1}
ant[{2, y_}, _, {0, 0}, _, _, {3, _}, _, _, _, _, _, _, _] :=
                                                        {RND, y + 1}
ant[{2, y_}, _, {0, 0}, _, _, _, {1, _}, _, _, _, _, _, _] :=
                                                        {RND, y + 1}
ant[{2, y_}, _, {0, 0}, _, _, _, _, _, _, _, {4, _}, _, _] :=
                                                        {RND, y + 1}
ant[{3, y_}, _, _, {0, 0}, _, _, {4, _}, _, _, _, _, _, _] :=
                                                        {RND, y + 1}
ant[{3, y_}, _, _, {0, 0}, _, _, _, {2, _}, _, _, _, _, _] :=
                                                        {RND, y + 1}
ant[{3, y_}, _, _, {0, 0}, _, _, _, _, _, _, _, {1, _}, _] :=
                                                        {RND, y + 1}
ant[{4, y_}, _, _, _, {0, 0}, _, _, {1, _}, _, _, _, _, _] :=
                                                        {RND, y + 1}
ant[{4, y_}, _, _, _, {0, 0}, _, _, _, {3, _}, _, _, _, _] :=
                                                        {RND, y + 1}
ant[{4, y_}, _, _, _, {0, 0}, _, _, _, _, _, _, _, {2, _}] :=
                                                        {RND, y + 1}
```

- An active ant facing an empty adjacent site vacates the site it is on.

```
ant[{1, _}, {0, 0}, _, _, _, _, _, _, _, _, _, _, _] := {0, 0}
ant[{2, _}, _, {0, 0}, _, _, _, _, _, _, _, _, _, _] := {0, 0}
ant[{3, _}, _, _, {0, 0}, _, _, _, _, _, _, _, _, _] := {0, 0}
ant[{4, _}, _, _, _, {0, 0}, _, _, _, _, _, _, _, _] := {0, 0}
```

For an empty site (a site having an ordered pair $\{0, 0\}$):

- An empty site faced by two or more active ants remains empty.

```
ant[{0, 0}, {3, _}, {4, _}, _, _, _, _, _, _, _, _, _, _] := {0, 0}
ant[{0, 0}, {3, _}, _, {1, _}, _, _, _, _, _, _, _, _, _] := {0, 0}
ant[{0, 0}, {3, _}, _, _, {2, _}, _, _, _, _, _, _, _, _] := {0, 0}
ant[{0, 0}, _, {4, _}, {1, _}, _, _, _, _, _, _, _, _, _] := {0, 0}
ant[{0, 0}, _, {4, _}, _, {2, _}, _, _, _, _, _, _, _, _] := {0, 0}
ant[{0, 0}, _, _, {1, _}, {2, _}, _, _, _, _, _, _, _, _] := {0, 0}
```

- An empty site faced by exactly one ant that has been active less than *s* time steps becomes occupied by the ant which randomly chooses a direction to face and increments its activity time clock by 1.

```
ant[{0, 0}, {3, y_?(# < s &)}, _, _, _, _, _, _, _, _, _, _] :=
                                                           {RND, y + 1}
ant[{0, 0}, _, {4, y_?(# < s &)}, _, _, _, _, _, _, _, _, _] :=
                                                           {RND, y + 1}
ant[{0, 0}, _, _, {1, y_?(# < s &)}, _, _, _, _, _, _, _, _] :=
                                                           {RND, y + 1}
ant[{0, 0}, _, _, _, {2, y_?(# < s &)}, _, _, _, _, _, _, _] :=
                                                           {RND, y + 1}
```

- An empty site remains unchanged.

```
ant[{0, 0}, _, _, _, _, _, _, _, _, _, _, _, _] := {0, 0}
```

- A border site remains unchanged.

```
ant[{b, 0}, _, _, _, _, _, _, _, _, _, _, _, _] := {b, 0}
```

Note: While it may be off-putting to see the large number of rules (38, to be exact), these rules are very simple to read and understand (which is helpful). Moreover, the use of rules in a CA program offers a substantial benefit: it is quite easy to add, delete, or modify the rules to obtain variations of the model (decreasing the time that we have to spend developing a program is quite important to us).

Applying the Rules

The following anonymous function is used to apply the ant rules to the 13 arguments representing a site and its neighbors:

```
MvonN[ant, #]&
```

where

```
MvonN[func__, lat_] :=
  MapThread[func, Map[RotateRight[lat, #]&,
            {{0, 0}, {1, 0}, {0, -1}, {-1, 0}, {0, 1},
             {1, -1}, {-1, -1}, {-1, 1}, {1, 1},
             {2, 0}, {0, -2}, {-2, 0}, {0, 2}}], 2]
```

The ant farm evolves over *t* time steps by repeatedly applying the anonymous update function to the antColony lattice representing the configuration of the ant colony using the NestList function.

```
evolve = NestList[MvonN[ant, #]&, initConf, t]
```

We can strip off the activity time clock from each lattice site by mapping the anonymous function

```
latticeConfig = Function[y, Map[#[[1]]&, y, {2}]]
```

onto the list given in evolve.

```
Map[latticeConfig, evolve]
```

The Ant Colony Program

```
maggiesFarm[n_, p_, s_, r_, t_]:=
Module[{antPopulation, antFarm, border, ant},

  antPopulation =
   Table[{{-1, 1, 2, 3, 4}[[Random[Integer, {1, 5}] ]],
          Random[Integer, {1, s}]}]* Floor[Random[] + p],
               {n - 1}, {n - 1}] /. {-1, _} -> {-1, 0};

  border =
   Append[Map[Append[#, {b, 0}]&, #], Table[{b, 0}, {Length[#] + 1}]]&;

  antFarm = border[antPopulation];

  RND := Random[Integer, {1, 4}];
```

```
ant[{-1,  0}, {x_?Positive, _}, _, _, _, _, _, _, _, _, _, _, _] :=
                                                     {RND, 1};
ant[{-1,  0},  _, {x_?Positive,_}, _, _, _, _, _, _, _, _, _, _] :=
                                                     {RND, 1};
ant[{-1,  0},  _, _, {x_?Positive,_}, _, _, _, _, _, _, _, _, _] :=
                                                     {RND, 1};
ant[{-1,  0},  _, _, _, {x_?Positive,_ , _, _, _, _, _, _, _, _] :=
                                                     {RND, 1};
ant[{-1, 0}, _, _, _, _, _, _, _, _, _, _, _] :=
          {-1 + # * Random[Integer, {2, 5}], #}&[Floor[Random[] + r]];
ant[{x_?Positive, s}, _, _, _, _, _, _, _, _, _, _, _, _] := {-1, 0};
ant[{1, y_}, {-1 | x_?Positive | b, _}, _, _, _, _,
                          _, _, _, _, _, _, _] := {RND, y + 1};
ant[{2, y_}, _, {-1 | x_?Positive | b, _}, _, _, _,
                          _, _, _, _, _, _, _] := {RND, y + 1};
ant[{3, y_}, _, _, {-1 | x_?Positive | b, _}, _, _,
                          _, _, _, _, _, _, _] := {RND, y + 1};
ant[{4, y_}, _, _, _, {-1 | x_?Positive | b, _}, _,
                          _, _, _, _, _, _, _] := {RND, y + 1};
ant[{1, y_}, {0, 0}, _, _, _, {4, _}, _, _, _, _, _, _, _] :=
                                                     {RND, y + 1};
ant[{1, y_}, {0, 0}, _, _, _, _, _, _, {2, _}, _, _, _, _] :=
                                                     {RND, y + 1};
ant[{1, y_}, {0, 0}, _, _, _, _, _, _, _, {3, _}, _, _, _] :=
                                                     {RND, y + 1};
ant[{2, y_}, _, {0, 0}, _, _, {3, _}, _, _, _, _, _, _, _] :=
                                                     {RND, y + 1};
ant[{2, y_}, _, {0, 0}, _, _, _, {1, _}, _, _, _, _, _, _] :=
                                                     {RND, y + 1};
ant[{2, y_}, _, {0, 0}, _, _, _, _, _, _, _, {4, _}, _, _] :=
                                                     {RND, y + 1};
ant[{3, y_}, _, _, {0, 0}, _, _, {4, _}, _, _, _, _, _, _] :=
                                                     {RND, y + 1};
ant[{3, y_}, _, _, {0, 0}, _, _, _, {2, _}, _, _, _, _, _] :=
                                                     {RND, y + 1};
ant[{3, y_}, _, _, {0, 0}, _, _, _, _, _, _, _, {1, _}, _] :=
                                                     {RND, y + 1};
ant[{4, y_}, _, _, _, {0, 0}, _, _, {1, _}, _, _, _, _, _] :=
                                                     {RND, y + 1};
ant[{4, y_}, _, _, _, {0, 0}, _, _, _, {3, _}, _, _, _, _] :=
                                                     {RND, y + 1};
ant[{4, y_}, _, _, _, {0, 0}, _, _, _, _, _, _, _, {2, _}] :=
                                                     {RND, y + 1};
ant[{1, _}, {0, 0}, _, _, _, _, _, _, _, _, _, _, _] := {0, 0};
ant[{2, _}, _, {0, 0}, _, _, _, _, _, _, _, _, _, _] := {0, 0};
```

```
ant[{3, _}, _, _, {0, 0}, _, _, _, _, _, _, _, _, _] := {0, 0};
ant[{4, _}, _, _, _, {0, 0}, _, _, _, _, _, _, _, _] := {0, 0};
ant[{0, 0}, {3, _}, {4, _}, _, _, _, _, _, _, _, _, _, _] := {0, 0};
ant[{0, 0}, {3, _}, _, {1, _}, _, _, _, _, _, _, _, _, _] := {0, 0};
ant[{0, 0}, {3, _}, _, _, {2, _}, _, _, _, _, _, _, _, _] := {0, 0};
ant[{0, 0}, _, {4, _}, {1, _}, _, _, _, _, _, _, _, _, _] := {0, 0};
ant[{0, 0}, _, {4, _}, _, {2, _}, _, _, _, _, _, _, _, _] := {0, 0};
ant[{0, 0}, _, _, {1, _}, {2, _}, _, _, _, _, _, _, _, _] := {0, 0};
ant[{0, 0}, {3, y_?(# < s &)}, _, _, _, _, _, _, _, _, _, _] :=
                                               {RND, y + 1};
ant[{0, 0}, _, {4, y_?(# < s &)}, _, _, _, _, _, _, _, _, _] :=
                                               {RND, y + 1};
ant[{0, 0}, _, _, {1, y_?(# < s &)}, _, _, _, _, _, _, _, _] :=
                                               {RND, y + 1};
ant[{0, 0}, _, _, _, {2, y_?(# < s &)}, _, _, _, _, _, _, _] :=
                                               {RND, y + 1};
ant[{0, 0}, _, _, _, _, _, _, _, _, _, _, _] := {0, 0};
ant[{b, 0}, _, _, _, _, _, _, _, _, _, _, _] := {b, 0};

MvonN[func__, lat_] :=
 MapThread[func, Map[RotateRight[lat, #]&,
          {{0, 0}, {1, 0}, {0, -1}, {-1, 0}, {0, 1},
           {1, -1}, {-1, -1}, {-1, 1}, {1, 1},
           {2, 0}, {0, -2}, {-2, 0}, {0, 2}}], 2];

NestList[MvonN[ant, #]&, antFarm, t];
]
```

"Go to the Ant, Thou Sluggard; Consider Her Ways"

The activity of the ant colony as a function of time is obtained by mapping an anonymous function that counts the number of active ants (given by the number of ordered pairs having a first component of 1, 2, 3 or 4) onto the matrices produced by the maggiesFarm program.

```
activityHistory = Map[Count[Flatten[#, 1], {x_?Positive, _}]&, evolve]
```

A plot of the ant colony activity over time is then obtained using

```
ListPlot[activityHistory]
```

Below we compare the behavior of an ant colony in which active ants become inactive after 10 time steps at a low density ($p = 0.1$) and high density ($p = 0.9$).

```
SeedRandom[8]
activityHistory =
   Map[Count[Flatten[#, 1], {x_?Positive, _}]&,
      Drop[maggiesFarm[10, 0.1, 10, 0.15, 250], 50]];
```

```
ListPlot[activityHistory,
        PlotJoined -> True,
        PlotRange -> {{0, 200}, {0, 10}},
        AxesLabel -> {FontForm["time step", {"Helvetica", 12}],
        FontForm["activity", {"Helvetica", 12}]}];
```

```
SeedRandom[8]
activityHistory =
   Map[Count[Flatten[#, 1], {x_?Positive, _}]&,
      Drop[maggiesFarm[10, 0.9, 10, 0.15, 250], 50]];
```

```
ListPlot[activityHistory,
        PlotJoined -> True,
        PlotRange -> {{0, 200}, {0, 100}},
        AxesLabel -> {FontForm["time step", {"Helvetica",12}],
        FontForm["activity", {"Helvetica", 12}]}];
```

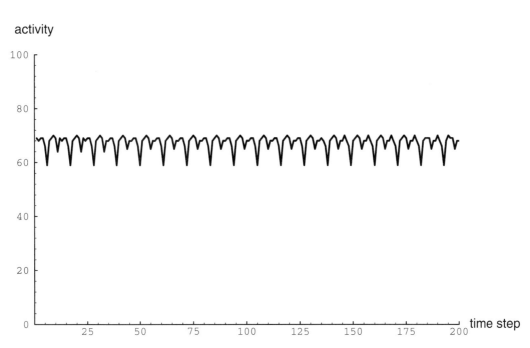

Comparing the two plots, we see that the chaotic activity pattern at low ant density is replaced by a periodic rhythm at high ant density.

The periodicity displayed by the ant colony is characteristic of an excitable medium system (see the Contagion chapter), which consists of spatially distributed elements that undergo an *excited-refractory-receptive* cycle, becoming excited as a result of interacting with neighboring elements, and subsequently returning incrementally to the quiescent state in which they are again receptive to being excited. In the ant colony model, the activated ant, $\{x\_Positive, 1\}$, represents the excited state, the active ant, $\{x\_Positive, y\_?(\# > 1\&)\}$, represents the refractory state, and the inactive ant, $\{-1, 0\}$, represents the receptive state.

Computer Simulation Project

Harvester ants are extremely industrious (hence the biblical quote from Solomon used in the title of the previous section) and they carry out a number of tasks. Four chores that take place outside the nest have been identified: foraging (retrieving food from a region around the nest), patrolling (responding to damage to the nest or invasion by alien ants and choosing the direction taken by foragers), nest maintenance (modifying the chambers of the nest and carrying away dirt from the nest entrance), and midden work (building and sorting the colony refuse pile and collecting and piling stones on the nest surface).

Each ant is able to carry out any of these tasks, and when performing a task each ant can be in either an active or inactive state. At a given time, the task performed by a particular ant (and hence the number of ants engaged in a particular task) depends on the number of ants that are engaged in the various tasks.

Create an ant colony CA where an ant determines at each time step which of eight states (inactive forager, active forager, inactive patroller, active patroller, inactive nest maintenance, active nest maintenance, inactive midden work, active midden work) to be in, using a two-step process:

First, an ant in a given task category (forager, patroller, nest maintenance, midden work) becomes or remains active (inactive) if the majority of the ants performing the same task in its immediate neighborhood (nearest neighbor sites), excluding itself, are inactive (active).

Second, an active ant switches from the task it is doing to another task if the majority of the active ants in its enlarged neighborhood (nearest neighbor and next nearest neighbor sites) are performing the same task as the ant.

References

Gaylord, Richard J. and Wellin, Paul R. *Computer Simulations with Mathematica: Explorations in Complex Physical and Biological Systems.* TELOS/Springer-Verlag (1994).

Goodwin, Brian. *How the Leopard Changed Its Spots.* Weidenfeld & Nicolson (1994). Chapter 3 "Life, an excitable medium."

Gordon, Deborah M. 1995. "The development of organization in an ant colony," American Scientist, 83 (Jan.-Feb.), 50–57.

Holldobler, Bert and Wilson, Edward O. *Journey to the Ants.* Belknap/Harvard University Press (1994).

Miramontes, Octavio, Sole, Ricard V., and Goodwin, Brian C. 1993. "Collective behavior of random-activated mobile cellular automata," Physica D, 63, 145–160.

14 Predator-Prey Ecosystems

Introduction

Ecology is concerned with the interrelationships between organisms and their environment. Observations of ecosystems containing two species, one of whom (known as the prey) serves as a food source for the other (known as the predator), indicate that the populations of the two species show out-of-phase oscillatory behavior. These cyclic fluctuations have been modeled analytically using coupled differential equations known as the Lotka-Volterra equations; however, it has been pointed out (Dewdney, 1984) that these equations "assume...a continuous predator continuously in search of a continuous prey." A discrete approach to modeling so-called predator-prey ecosystems employs a lattice system that represents territory that predators and prey co-habitate. Some of the lattice sites are empty and other sites are occupied by a predator or a prey. As the system evolves over a number of time steps, the predators and/or prey execute random walks on the lattice in the course of which predators eat prey, prey are eaten by predators, predators and preys give birth, and predators starve to death. We will develop a simple cellular automaton version of the predator-prey model that exhibits out-of-phase oscillatory fluctuations in the populations of the predator and prey species.

The Grazing Herd CA

In the grazing herd cellular automaton, the prey (the food source) remain stationary while the predators move about, and new food sources appear randomly in vacant places in each time step. This model might be used to represent a herd of animals (e.g., cows or sheep) grazing on the vegetation (e.g., grass) in a pasture.

The System

The predator-prey system employs a two-dimensional square lattice with periodic boundary conditions. Each lattice site is either empty or is occupied by a predator or a prey.

The lattice sites that are empty have value 0 and the lattice sites that contain a food source (i.e., that are occupied by a prey) have value 1. The value of a lattice site occupied by a predator is a tuplet, $\{x, y, z\}$.

The first component of the tuplet is an integer value 1, 2, 3, or 4 indicating the direction faced by the predator. The value 1 indicates a site occupied by a north-facing predator, the value 2 indicates a site occupied by an east-facing predator, the value 3 indicates a site occupied by a south-facing predator, and the value 4 indicates a site occupied by a west-facing predator. The value of the first component is randomly chosen in each time step.

The second component of the tuplet is a nonnegative integer value indicating the amount of time that remains until a newborn predator is due to be born. At the moment that a predator gives birth, the value of the birth time clock is set equal to "preg" (this is the normal gestation period for the predator species). At each ensuing time step, the value of the birth clock decreases by 1 unless its value is 0, in which case it remains 0. When the birth clock value is 0, the baby is born when the predator moves but not while it remains in place.

The third component of the tuplet is a nonnegative integer value indicating the amount of time that remains before a predator starves to death. At the moment that a predator eats a prey, its diet time clock is set equal to "starve." At each ensuing time step, the value of the diet clock decreases by 1. When the diet clock value is 0, the predator dies and vanishes from the lattice.

Initially, predators and prey are randomly placed on the lattice with probability predDensity and preyDensity, respectively (in the large system limit, the densities of predators and prey will be given by these quantities), using

```
pasture =
  Table[Floor[Random[] + (preyDensity + predDensity)] *
       Floor[1 + Random[] + predDensity/(preyDensity + predDensity)],
       {n}, {n}] /.
  2 :> {RND, Random[Integer, {0, preg}], Random[Integer, {1, starve}]}
```

where

```
RND := Random[Integer, {1, 4}]
```

Each predator in `pasture` faces a randomly chosen direction, its birth clock value is randomly set between 0 and `preg`, and its diet clock value is randomly set between 0 and `starve`.

The Update Rules

The rules governing the behaviors of the predators and prey are implemented as rewrite rules for updating the values of the sites of the CA lattice. These rules take 13 arguments in the following order:

```
eco[site, N, E, S, W, NE, SE, SW, NW, Nn, Ee, Ss, Ww]
```

where the 13 arguments represent the value of the site, the values of the four nearest neighbors in the N, E, S, W directions, the values of the four nearest neighbors in the NE, SE, SW, NW directions, and the values of four next nearest neighbors in the N, E, S, W directions.

The 53 rewrite rules are as follows:

- A predator with 0 birth and diet clock values starves to death and vacates the site, leaving behind a newborn predator facing a randomly chosen direction with a birth clock value of `preg` and a diet clock value of `starve`.

```
eco[{_, 0, 0}, _, _, _, _, _, _, _, _, _, _, _, _] :=
                                          {RND, preg, starve}
```

- A predator with a 0 diet clock value and a nonzero birth clock value starves to death and vanishes, leaving an empty site.

```
eco[{_, _, 0}, _, _, _, _, _, _, _, _, _, _, _, _] := 0
```

- A predator with a positive diet clock value that faces an adjacent empty or food site that is faced by at least one other predator with a positive diet clock value, remains in place, randomly chooses a direction to face, decrements its diet clock by 1, and decrements its birth clock by one unless its value is 0.

```
eco[{1, a_, b_?Positive}, 0 | 1, _, _, _, {4, _, _?Positive},
                    _, _, _, _, _, _, _] := {RND, Max[0, a - 1], b - 1}
eco[{1, a_, b_?Positive}, 0 | 1, _, _, _, _, _, _,
          {2, _, _?Positive}, _, _, _, _] := {RND, Max[0, a - 1], b - 1}
eco[{1, a_, b_?Positive}, 0 | 1, _, _, _, _, _, _,
            {3, _, _?Positive}, _, _, _] := {RND, Max[0, a - 1], b - 1}
```

```
eco[{2, a_, b_?Positive}, _, 0 | 1, _, _, {3, _, _?Positive},
                _, _, _, _, _, _, _] := {RND, Max[0, a - 1], b - 1}
eco[{2, a_, b_?Positive}, _, 0 | 1, _, _, _, {1, _, _?Positive},
                _, _, _, _, _, _] := {RND, Max[0, a - 1], b - 1}
eco[{2, a_, b_?Positive}, _, 0 | 1, _, _, _, _, _
        _, _, {4, _, _?Positive}, _, _] := {RND, Max[0, a - 1], b - 1}
eco[{3, a_, b_?Positive}, _, _, 0 | 1, _, _, {4, _, _?Positive},
                _, _, _, _, _, _] := {RND, Max[0, a - 1], b - 1}
eco[{3, a_, b_?Positive}, _, _, 0 | 1, _, _, _
    {2, _, _?Positive}, _, _, _, _, _] := {RND, Max[0, a - 1], b - 1}
eco[{3, a_, b_?Positive}, _, _, 0 | 1, _, _, _, _
        _, _, _, {1, _, _?Positive}, _] := {RND, Max[0, a - 1], b - 1}
eco[{4, a_, b_?Positive}, _, _, _, 0 | 1, _, _
    {1, _, _?Positive}, _, _, _, _, _] := {RND, Max[0, a - 1], b - 1}
eco[{4, a_, b_?Positive}, _, _, _, 0 | 1, _, _, _
    {3, _, _?Positive}, _, _, _, _] := {RND, Max[0, a - 1], b - 1}
eco[{4, a_, b_?Positive}, _, _, _, 0 | 1, _, _, _
        _, _, _, _, {2, _, _?Positive}] := {RND, Max[0, a - 1], b - 1}
```

- A prey site that is faced by two or more predators with positive diet
  clock values on adjacent sites remains unchanged.

```
eco[1, {3, _, _?Positive}, {4, _, _?Positive},
                        _, _, _, _, _, _, _, _, _] := 1
eco[1, {3, _, _?Positive}, _, {1, _, _?Positive},
                        _, _, _, _, _, _, _, _] := 1
eco[1, {3, _, _?Positive}, _, _, {2, _, _?Positive},
                        _, _, _, _, _, _, _, _] := 1
eco[1, _, {4, _, _?Positive}, {1, _, _?Positive},
                        _, _, _, _, _, _, _, _] := 1
eco[1, _, {4, _, _?Positive}, _, {2, _, _?Positive},
                        _, _, _, _, _, _, _] := 1
eco[1, _, _, {1, _, _?Positive}, {2, _, _?Positive},
                        _, _, _, _, _, _, _, _] := 1
```

- An empty site that is faced by two or more predators with positive
  diet clock values becomes a food source with probability p.

```
eco[0, {3, _, _?Positive}, {4, _, _?Positive},
                _, _, _, _, _, _, _, _, _] := Floor[p + Random[]]
eco[0, {3, _, _?Positive}, _, {1, _, _?Positive},
                _, _, _, _, _, _, _, _] := Floor[p + Random[]]
eco[0, {3, _, _?Positive}, _, _, {2, _, _?Positive},
                _, _, _, _, _, _, _] := Floor[p + Random[]]
```

```
eco[0, _, {4, _, _?Positive}, {1, _, _?Positive},
                    _, _, _, _, _, _, _, _, _] := Floor[p + Random[]]
eco[0, _, {4, _, _?Positive}, _, {2, _, _?Positive},
                    _, _, _, _, _, _, _, _] := Floor[p + Random[]]
eco[0, _, _, {1, _, _?Positive}, {2, _, _?Positive},
                    _, _, _, _, _, _, _, _] := Floor[p + Random[]]
```

- A predator with a 0 birth clock value and a positive diet clock value that faces an adjacent empty or prey site that is faced by no other predator with a positive diet clock value, vacates the site it is occupying, leaving behind a newborn predator facing a randomly chosen direction with a birth clock value of preg and a diet clock value of starve.

```
eco[{1, 0, _?Positive}, 0 | 1, _, _, _, _, _, _, _, _, _, _, _] :=
                                                  {RND, preg, starve}
eco[{2, 0, _?Positive}, _, 0 | 1, _, _, _, _, _, _, _, _, _, _] :=
                                                  {RND, preg, starve}
eco[{3, 0, _?Positive}, _, _, 0 | 1, _, _, _, _, _, _, _, _, _] :=
                                                  {RND, preg, starve}
eco[{4, 0, _?Positive}, _, _, _, 0 | 1, _, _, _, _, _, _, _, _] :=
                                                  {RND, preg, starve}
```

- A predator with positive birth and diet clock values that faces an adjacent empty or prey site that is faced by no other predator with a positive diet clock value, vacates the site it is occupying, leaving it empty.

```
eco[{1, _, _?Positive}, 0 | 1, _, _, _, _, _, _, _, _, _, _, _] := 0
eco[{2, _, _?Positive}, _, 0 | 1, _, _, _, _, _, _, _, _, _, _] := 0
eco[{3, _, _?Positive}, _, _, 0 | 1, _, _, _, _, _, _, _, _, _] := 0
eco[{4, _, _?Positive}, _, _, _, 0 | 1, _, _, _, _, _, _, _, _] := 0
```

- A prey site that is faced by exactly one predator with a 0 birth clock value and a positive diet clock value on an adjacent site, is occupied by the predator, which randomly chooses a direction to face, resets its birth clock to preg, and resets its diet clock to starve.

```
eco[1, {3, 0, _?Positive}, _, _, _, _, _, _, _, _, _, _, _] :=
                                                  {RND, preg, starve}
eco[1, _, {4, 0, _?Positive}, _, _, _, _, _, _, _, _, _, _] :=
                                                  {RND, preg, starve}
eco[1, _, _, {1, 0, _?Positive}, _, _, _, _, _, _, _, _, _] :=
                                                  {RND, preg, starve}
eco[1, _, _, _, {2, 0, _?Positive}, _, _, _, _, _, _, _, _] :=
                                                  {RND, preg, starve}
```

- An empty site that is faced by exactly one predator with a 0 birth clock value and a positive diet clock value on an adjacent site is occupied by the predator, which randomly chooses a direction to face, decrements its diet clock by 1, and resets its birth clock to preg.

```
eco[0, {3, 0, b_?Positive}, _, _, _, _, _, _, _, _, _, _] :=
                                               {RND, preg, b - 1}
eco[0, _, {4, 0, b_?Positive}, _, _, _, _, _, _, _, _, _] :=
                                               {RND, preg, b - 1}
eco[0, _, _, {1, 0, b_?Positive}, _, _, _, _, _, _, _, _] :=
                                               {RND, preg, b - 1}
eco[0, _, _, _, {2, 0, b_?Positive}, _, _, _, _, _, _, _] :=
                                               {RND, preg, b - 1}
```

- A prey site that is faced by exactly one predator with positive birth and diet clock values on an adjacent site is occupied by the predator, which randomly selects a direction to face, decrements its birth clock by 1, and resets its diet clock to starve.

```
eco[1, {3, a_, _?Positive}, _, _, _, _, _, _, _, _, _, _] :=
                                               {RND, a -1, starve}
eco[1, _, {4, a_, _?Positive}, _, _, _, _, _, _, _, _, _] :=
                                               {RND, a -1, starve}
eco[1, _, _, {1, a_, _?Positive}, _, _, _, _, _, _, _, _] :=
                                               {RND, a -1, starve}
eco[1, _, _, _, {2, a_, _?Positive}, _, _, _, _, _, _, _] :=
                                               {RND, a -1, starve}
```

- An empty site that is faced by exactly one predator with positive diet and birth clock values on an adjacent site is occupied by the predator, which randomly selects a direction to face and decrements its diet and birth clocks by 1.

```
eco[0, {3, a_, b_?Positive}, _, _, _, _, _, _, _, _, _, _] :=
                                               {RND, a - 1, b - 1}
eco[0, _, {4, a_, b_?Positive}, _, _, _, _, _, _, _, _, _] :=
                                               {RND, a - 1, b - 1}
eco[0, _, _, {1, a_, b_?Positive}, _, _, _, _, _, _, _, _] :=
                                               {RND, a - 1, b - 1}
eco[0, _, _, _, {2, a_, b_?Positive}, _, _, _, _, _, _, _] :=
                                               {RND, a - 1, b - 1}
```

- Any other predator remains in place, randomly chooses a direction to face, and decrements its diet and birth time clocks by 1.

```
eco[{_, a_, b_}, _, _, _, _, _, _, _, _, _, _, _, _] :=
                                           {RND, a - 1, b - 1}
```

- Any other prey site remains unchanged.

```
eco[1, _, _, _, _, _, _, _, _, _, _, _, _] := 1
```

- Any other empty site becomes a food source with probability *p*.

```
eco[0, _, _, _, _, _, _, _, _, _, _, _, _] := Floor[p + Random[]]
```

Applying the Rules

The sites in the lattice are updated at each time step by applying the
following anonymous function to the CA lattice:

```
MvonN[eco, #]&
```

where

```
MvonN[func__, lat_] :=
  MapThread[func, Map[RotateRight[lat, #]&,
            {{0, 0}, {1, 0}, {0, -1}, {-1, 0}, {0, 1},
             {1, -1}, {-1, -1}, {-1, 1}, {1, 1},
             {2, 0}, {0, -2}, {-2, 0}, {0, 2}}], 2]
```

The pasture of predators and prey evolves over *t* time steps using the
following Nest operation.

```
NestList[MvonN[eco, #]&, pasture, t]
```

The Program

```
PredatorPrey[n_, preyDensity_, predDensity_, preg_,
            starve_, p_, t_] /; preyDensity + predDensity < 1 :=
 Module[{eco, RND, pasture MvonN},

 RND:=Random[Integer,{1,4}];

 pasture =
  Table[Floor[Random[] + (preyDensity + predDensity)] *
        Floor[1 + Random[] + predDensity/(preyDensity + predDensity)],
        {n}, {n}] /.
 2 :> {RND, Random[Integer, {0, preg}], Random[Integer, {1, starve}]}];
```

```
eco[{_, 0, 0}, _, _, _, _, _, _, _, _, _, _, _, _] :=
                                            {RND, preg, starve};
eco[{_, _, 0}, _, _, _, _, _, _, _, _, _, _, _, _] := 0;
eco[{1, a_, b_?Positive}, 0 | 1, _, _, _, {4, _, _?Positive},
              _, _, _, _, _, _, _] := {RND, Max[0, a - 1], b - 1};
eco[{1, a_, b_?Positive}, 0 | 1, _, _, _, _, _, _
    {2, _, _?Positive}, _, _, _, _] := {RND, Max[0, a - 1], b - 1};
eco[{1, a_, b_?Positive}, 0 | 1, _, _, _, _, _, _, _
        {3, _, _?Positive}, _, _, _] := {RND, Max[0, a - 1], b - 1};
eco[{2, a_, b_?Positive}, _, 0 | 1, _, _, {3, _, _?Positive},
              _, _, _, _, _, _, _] := {RND, Max[0, a - 1], b - 1};
eco[{2, a_, b_?Positive}, _, 0 | 1, _, _, _, {1, _, _?Positive},
                 _, _, _, _, _, _] := {RND, Max[0, a - 1], b - 1};
eco[{2, a_, b_?Positive}, _, 0 | 1, _, _, _, _, _
       _, _, {4, _, _?Positive}, _, _] := {RND, Max[0, a - 1], b - 1};
eco[{3, a_, b_?Positive}, _, _, 0 | 1, _, _, {4, _, _?Positive},
                 _, _, _, _, _, _] := {RND, Max[0, a - 1], b - 1};
eco[{3, a_, b_?Positive}, _, _, 0 | 1, _, _, _
    {2, _, _?Positive}, _, _, _, _, _] := {RND, Max[0, a - 1], b - 1};
eco[{3, a_, b_?Positive}, _, _, 0 | 1, _, _, _, _
       _, _, _, {1, _, _?Positive}, _] := {RND, Max[0, a - 1], b - 1};
eco[{4, a_, b_?Positive}, _, _, _, 0 | 1, _, _
    {1, _, _?Positive}, _, _, _, _, _] := {RND, Max[0, a - 1], b - 1};
eco[{4, a_, b_?Positive}, _, _, _, 0 | 1, _, _, _
        {3, _, _?Positive}, _, _, _, _] := {RND, Max[0, a - 1], b - 1};
eco[{4, a_, b_?Positive}, _, _, _, 0 | 1, _, _, _
       _, _, _, _, {2, _, _?Positive}] := {RND, Max[0, a - 1], b - 1};
eco[1, {3, _, _?Positive}, {4, _, _?Positive},
                 _, _, _, _, _, _, _, _, _, _] := 1;
eco[1, {3, _, _?Positive}, _, {1, _, _?Positive},
                 _, _, _, _, _, _, _, _, _] := 1;
eco[1, {3, _, _?Positive}, _, _, {2, _, _?Positive},
                 _, _, _, _, _, _, _, _] := 1;
eco[1, _, {4, _, _?Positive}, {1, _, _?Positive},
                 _, _, _, _, _, _, _, _, _] := 1;
eco[1, _, {4, _, _?Positive}, _, {2, _, _?Positive},
                 _, _, _, _, _, _, _, _] := 1;
eco[1, _, _, {1, _, _?Positive}, {2, _, _?Positive},
                 _, _, _, _, _, _, _, _] := 1;
eco[0, {3, _, _?Positive}, {4, _, _?Positive},
                 _, _, _, _, _, _, _, _, _, _] := Floor[p + Random[]];
eco[0, {3, _, _?Positive}, _, {1, _, _?Positive},
                 _, _, _, _, _, _, _, _, _] := Floor[p + Random[]];
eco[0, {3, _, _?Positive}, _, _, {2, _, _?Positive},
                 _, _, _, _, _, _, _, _] := Floor[p + Random[]];
```

```
eco[0, _, {4, _, _?Positive}, {1, _, _?Positive},
                _, _, _, _, _, _, _, _, _] := Floor[p + Random[]];
eco[0, _, {4, _, _?Positive}, _, {2, _, _?Positive},
                _, _, _, _, _, _, _, _] := Floor[p + Random[]];
eco[0, _, _, {1, _, _?Positive}, {2, _, _?Positive},
                _, _, _, _, _, _, _, _] := Floor[p + Random[]];
eco[{1, 0, _?Positive}, 0 | 1, _, _, _, _, _, _, _, _, _, _] :=
                                        {RND, preg, starve};
eco[{2, 0, _?Positive}, _, 0 | 1, _, _, _, _, _, _, _, _, _] :=
                                        {RND, preg, starve};
eco[{3, 0, _?Positive}, _, _, 0 | 1, _, _, _, _, _, _, _, _] :=
                                        {RND, preg, starve};
eco[{4, 0, _?Positive}, _, _, _, 0 | 1, _, _, _, _, _, _, _] :=
                                        {RND, preg, starve};
eco[{1, _, _?Positive}, 0 | 1, _, _, _, _, _, _, _, _, _, _] := 0;
eco[{2, _, _?Positive}, _, 0 | 1, _, _, _, _, _, _, _, _, _] := 0;
eco[{3, _, _?Positive}, _, _, 0 | 1, _, _, _, _, _, _, _, _] := 0;
eco[{4, _, _?Positive}, _, _, _, 0 | 1, _, _, _, _, _, _, _] := 0;
eco[1, {3, 0, _?Positive}, _, _, _, _, _, _, _, _, _, _] :=
                                        {RND, preg, starve};
eco[1, _, {4, 0, _?Positive}, _, _, _, _, _, _, _, _, _] :=
                                        {RND, preg, starve};
eco[1, _, _, {1, 0, _?Positive}, _, _, _, _, _, _, _, _] :=
                                        {RND, preg, starve};
eco[1, _, _, _, {2, 0, _?Positive}, _, _, _, _, _, _, _] :=
                                        {RND, preg, starve};
eco[0, {3, 0, b_?Positive}, _, _, _, _, _, _, _, _, _, _] :=
                                        {RND, preg, b - 1};
eco[0, _, {4, 0, b_?Positive}, _, _, _, _, _, _, _, _, _] :=
                                        {RND, preg, b - 1};
eco[0, _, _, {1, 0, b_?Positive}, _, _, _, _, _, _, _, _] :=
                                        {RND, preg, b - 1};
eco[0, _, _, _, {2, 0, b_?Positive}, _, _, _, _, _, _, _] :=
                                        {RND, preg, b - 1};
eco[1, {3, a_, _?Positive}, _, _, _, _, _, _, _, _, _, _] :=
                                        {RND, a -1, starve};
eco[1, _, {4, a_, _?Positive}, _, _, _, _, _, _, _, _, _] :=
                                        {RND, a -1, starve};
eco[1, _, _, {1, a_, _?Positive}, _, _, _, _, _, _, _, _] :=
                                        {RND, a -1, starve};
eco[1, _, _, _, {2, a_, _?Positive}, _, _, _, _, _, _, _] :=
                                        {RND, a -1, starve};
eco[0, {3, a_, b_?Positive}, _, _, _, _, _, _, _, _, _, _] :=
                                        {RND, a - 1, b - 1};
```

```
eco[0, _, {4, a_, b_?Positive}, _, _, _, _, _, _, _, _, _, _] :=
                                              {RND, a - 1, b - 1};
eco[0, _, _, {1, a_, b_?Positive}, _, _, _, _, _, _, _, _, _] :=
                                              {RND, a - 1, b - 1};
eco[0, _, _, _, {2, a_, b_?Positive}, _, _, _, _, _, _, _, _] :=
                                              {RND, a - 1, b - 1};
eco[{_, a_, b_}, _, _, _, _, _, _, _, _, _, _, _, _] :=
                                              {RND, a - 1, b - 1};
eco[1, _, _, _, _, _, _, _, _, _, _, _, _] := 1;
eco[0, _, _, _, _, _, _, _, _, _, _, _, _] := Floor[p + Random[]];


MvonN[func__, lat_] :=
 MapThread[func, Map[RotateRight[lat, #]&,
            {{0, 0}, {1, 0}, {0, -1}, {-1, 0}, {0, 1},
            {1, -1}, {-1, -1}, {-1, 1}, {1, 1},
            {2, 0}, {0, -2}, {-2, 0}, {0, 2}}], 2];


NestList[MvonN[eco, #]&, pasture, t]
]
```

Running the Program

We can run the `PredatorPrey` program and then calculate the number of predators and prey as a function of time.

```
SeedRandom[2]
graze = PredatorPrey[20, 0.2, 0.1, 2, 3, 0.1, 100];

predatorPop = Map[Count[Flatten[#, 1], {__}]&, graze];

preyPop = Map[Count[Flatten[#, 1], 1]&, graze];
```

We can look at the oscillatory behavior in the numbers of predators and prey over time by dropping the initial 20 time steps in `predatorPop` and `preyPop`,

```
PredPreyPop = Map[Drop[#, 20]&, {predatorPop, preyPop}];
```

pairing off each of the remaining population values with its corresponding time step value,

```
PredPreyPopTimePairs =
               Map[Transpose[{Range[20, 100], #}]&, PredPreyPop];
```

and using the `MultipleListPlot` function in the *Mathematica* Graphics package.

```
Needs["Graphics'MultipleListPlot'"]


MultipleListPlot[ PredPreyPopTimePairs /. {x__} -> x,
    PlotLabel  -> FontForm["GRAZING HERD ECOSYSTEM", {"Palatino", 14}],
    AxesOrigin -> {0,  Min[PredPreyPop]  - 10},
    AxesLabel  -> {FontForm["time", {"NewYork", 12}],
                   FontForm["population", {"NewYork", 12}]},
    PlotRange  -> {{0, 120}, {Min[#] - 10, Max[#] + 10}&[PredPreyPop]},
    PlotJoined -> True]
```

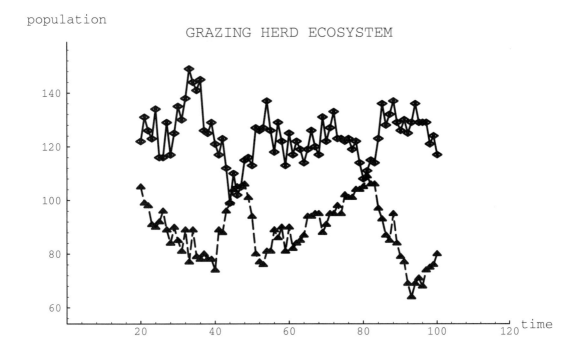

-Graphics-

As the figure shows, the predators and prey display oscillatory population density fluctuations which are out-of-phase with one another. This cyclic behavior isn't surprising. The predators (e.g., cows or sheep) eat the prey (e.g., grass) and reproduce, increasing the size of the herd while depleting the food source supply. When the food source is sufficiently depleted, the predators start to die of starvation and therefore breed less. When the size of the herd is sufficently diminished, the growth of new food sources begins to increase, and so on.

Computer Simulation Project

Our predator-prey CA, which assumes that prey remain stationary while predators move around, is appropriate for an animal-vegetation ecosystem. For an animal-animal ecosystem, on the other hand, both the predator and prey might be mobile (see for example, the Wa-Tor model (Dewdney, 1984)). Modify the grazing herd CA to allow for mobile prey which gestate.

References

Dewdney, A. K. 1984. "Sharks and fish wage an ecological war on the toroidal planet Wa-Tor." Scientific American, 251(12), 14–22.

Mitchel Resnick. *Turtles, Termites, and Traffic Jams*. MIT Press (1994), pp. 88–94.

15 | Contagion in Excitable Media

Introduction

A number of physical, chemical, biological, and social phenomena have been modeled using excitable media models, including the formation of stars in spiral disk galaxies, oscillatory chemical reactions, the contraction of cardiac tissue, and infectious disease epidemics.

An excitable medium consists of spatially distributed elements that undergo an *excited-refractory-receptive* cycle. In this cycle, an element in the quiescent state enters the excited state as a result of interacting with an excited neighboring element and then returns incrementally to the quiescent state in which it is again receptive to being excited.

The process by which elements go from the quiescent state to the excited state is a form of reversible contagious spreading (of the excited state) since an excited element does not remain indefinitely in that state (in contrast to the irreversible spreading that occurs when excited elements remain excited).

The excited-refractory-receptive cycle that characterizes excitable media systems can be modeled using multi-state cellular automata. We will look at both deterministic and probabilistic versions of the excitable media CA.

The Deterministic Epidemic CA

In a deterministic excitable medium CA, an element remains in any given state either for a definite amount of time or until some (nonprobabilistic) condition is met and it then goes to another state. The epidemic model (Schonfisch, 1995) employs a system (representing a population) consisting of an *n*-by-*n* square lattice with periodic boundary conditions. Each lattice site is occupied by an individual who is either *susceptible*, *infectious*, or *immune*.

Each individual is represented by a non-negative integer value. The following integer values are used to indicate the state of an individual:

0 — a susceptible individual

$1, 2, \ldots, a$ — an infectious individual

$(a + 1), (a + 2), \ldots, (a + g)$ — an immune individual

Initially, the system contains (in the large system limit) a density s of individuals in various stages of infectiousness and immunity, randomly distributed among susceptible individuals.

```
population =
 Table[Floor[1 + s - Random[] ]*Random[Integer, {1, a + g}], {n}, {n}]
```

The values of the lattice sites in the population are updated in each time step using rules of the form spread[$site, N, E, S, W$], where the five arguments are the values of a site and its nearest neighbor sites in the north, east, south, and west directions.

The update rules are

- A susceptible individual becomes infectious if at least one nearest neighbor site is infectious.

```
spread[0, u_, v_, w_, x_] := 1 /; MemberQ[Range[a], u | v | w | x]
```

- An individual that has been immune for g time steps becomes susceptible, and a susceptible individual that has no infectious nearest neighbors remains susceptible.

```
spread[ 0 | (a + g), _, _, _, _ ] := 0
```

- An individual that is either infectious or has been immune for less than g time steps increments its value.

```
spread[x_?Positive, _, _, _, _] := x + 1
```

Note: According to this rule, an individual that has been infectious for a time steps becomes immune.

The spread rules can be applied to the CA lattice using the NestList function with the following anonymous function

```
VonNeumann[spread, #]&
```

where

```
VonNeumann[func__, lat_] :=
    MapThread[func, Map[RotateRight[lat, #]&,
              {{0, 0}, {1, 0}, {0, -1}, {-1, 0}, {0, 1}}], 2]
```

The program is constructed from these code fragments.

The Program

```
contagion[n_, s_, a_, g_, t_] :=
 Module[{population, spread, VonNeumann},

  population = Table[Floor[1 + s - Random[]] *
                     Random[Integer, {1, a + g}] , {n}, {n}];

  spread[ 0 | (a + g), _, _, _, _ ] := 0;
  spread[x_?Positive, _, _, _, _] := x + 1;
  spread[0, u_, v_, w_, x_] := 1 /; MemberQ[Range[a], u | v | w | x];

  VonNeumann[func__, lat_] :=
    MapThread[func, Map[RotateRight[lat, #]&,
              {{0, 0}, {1, 0}, {0, -1}, {-1, 0}, {0, 1}}], 2];

  NestList[VonNeumann[spread, #]&, population, t]
 ]
```

Running the Program

When a deterministic excitable medium CA is run with suitably chosen parameter values, various spatially distributed patterns emerge over time. This can be illustrated with the contagion CA.

```
SeedRandom[1]
Show[GraphicsArray[Partition[
  Map[Show[Graphics[RasterArray[# /.
            Thread[Range[0, Max[#]] ->
                          Map[Hue, Table[Random[], {Max[#] + 1}]]]],
          DisplayFunction -> Identity, AspectRatio -> Automatic] ]&,
  contagion[100, 0.05, 8, 8, 200][[{1, 5, 10, 50, 100, 200}]] ], 2] ]];
```

The Stochastic Forest Fire CA

The forest fire model (Dossel and Schwabl, 1994) uses a system consisting of an n-by-n square lattice with periodic boundary conditions. The values of the lattice sites are 0, 1, or 2 where

0 — an empty site
1 — a site occupied by a tree
2 — a site occupied by a burning tree

Initially, the system contains of trees and burning trees randomly distributed among empty sites.

```
forestPreserve = Table[Floor[1 + s - Random[]], {n}, {n}] /.
                                  1 :> Floor[1 + k + Random[]]
```

where s is the probability of a lattice site being occupied by a tree (in the large system limit, s is the density of trees) and k is the probability that a tree is burning (in the large system limit, k is the fraction of trees that are burning).

There are four rules for updating the CA lattice sites. The five arguments of the rules are the values of the sites in the von Neumann neighborhood of a given site (i.e., the values of a site and its nearest neighbor sites in the north, east, south, and west directions, in that order).

Three of the rules are probabilistic in the sense that the outcome of applying the rule depends on generating a random number. Three probability parameters are used:

p — the tree growth probability
f — the lightning probability
g — the immunity probability

During a time step, all of the sites are updated according to the following rules:

- An empty site sprouts a tree $(0 \rightarrow 1)$ with probability p.

```
spread[0, _, _, _, _] := Floor[1 + p - Random[] ]
```

- A tree catches fire $(1 \rightarrow 2)$ with probability $(1 - g)$ if at least one nearest neighbor tree is burning.

```
spread[1, a_, b_, c_, d_ ] := 1 + Floor[1 + (1 - g) - Random[] ] /;
                              MatchQ[2, a | b | c | d]
```

- A tree catches fire ($1 \rightarrow 2$) with probability $f * (1 - g)$ if no nearest neighbor tree is burning.

```
spread[1, a_, b_, c_, d_ ] := 1 + Floor[1 + f (1 - g) - Random[]]
```

- A burning tree burns down and becomes an empty site ($2 \rightarrow 0$).

```
spread[2, _, _, _, _] = 0
```

The rules can be applied to the CA lattice by using the `NestList` function with the following anonymous function:

```
VonNeumann[spread, #]&
```

where

```
VonNeumann[func__, lat_] :=
    MapThread[func, Map[RotateRight[lat, #]&,
              {{0, 0}, {1, 0}, {0, -1}, {-1, 0}, {0, 1}}], 2]
```

The program is constructed from these code fragments.

The Program

The input parameters of the program are, in order, lattice size, initial tree probability, initial burning tree probability, tree birth probability, lightning strike probability, immunity probability, and the number of time steps.

```
forestFire[n_, s_, k_, p_, f_, g_, t_] :=
Module[{initConf, spread, sprout, catch, spont},

  sprout = (1 + p);
  catch = (2 - g);
  spont = 1 + f (1- g);

  forestPreserve = Table[Floor[1 + s - Random[]], {n}, {n}] /.
                                  1 :> Floor[1 + k + Random[]];

  spread[0, _, _, _, _] := Floor[sprout - Random[] ];
  spread[2, _, _, _, _] = 0;
  spread[1, a_, b_, c_, d_] :=
          1 + Floor[catch - Random[] ] /; MatchQ[2, a | b | c | d];
  spread[1, a_, b_, c_, d_] := 1 + Floor[spont - Random[]];
```

```
VonNeumann[func__, lat_] :=
    MapThread[func, Map[RotateRight[lat, #]&,
                 {{0, 0}, {1, 0}, {0, -1}, {-1, 0}, {0, 1}}], 2];

    NestList[VonNeumann[spread, #]&, forestPreserve, t]
]
```

Note: We define the three quantities—sprout, catch, and spont—in the forestFire program so that the quantities, $(1+p)$, $(2-g)$, and $1+f(1-g)$, don't have to be recalculated each time the spread rules in which these quantities appear is applied.

Running the Program

Three regimes of behavior have been found (Dossel and Schwabl, 1994) for the forest fire model. These are shown below.

(i) spiral-shaped fire fronts

```
SeedRandom[2]
Show[Graphics[RasterArray[
            Reverse[forestFire[150, 0.3, 0.001, 0.05, 0, 0, 100] ] /.
                Thread[{0, 1, 2} -> {RGBColor[0.6, 0.2, 0.1],
                                     RGBColor[0.1, 0.75, 0.2],
                                     RGBColor[1, 1, 0]} ] ] ] ],
        AspectRatio -> Automatic];
```

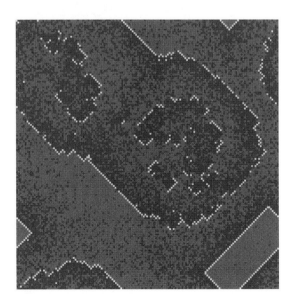

When $f = g = 0$ and $p \ll 1$, clusters of isolated forests of (non-burning) trees emerge within empty regions and these clusters persist until they grow into other clusters that are burning. The fire fronts separate the empty and forested areas.

(ii) self-organized critical (SOC) state

```
SeedRandom[1]
Show[Graphics[RasterArray[
        Reverse[forestFire[150, 0.3, 0,  0.05, 0.00025, 0, 500]] /.
            Thread[{0, 1, 2} -> {RGBColor[0.6, 0.2, 0.1],
                                 RGBColor[0.1, 0.75, 0.2],
                                 RGBColor[1, 1, 0]}]]],
        AspectRatio -> Automatic];
```

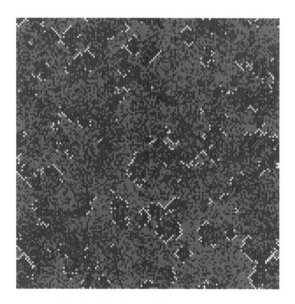

When $g = 0$, $p \ll 1$, and $f \ll p$, two time scale separations occur: tree growth occurs much more frequently than lightning strikes while forest clusters burn down faster than they grow. This results in an SOC state in which clusters of forests of all sizes are burning.

(iii) percolation transition

```
Show[Graphics[RasterArray[
            Reverse[forestFire[150, 0.3, 0, 0.2, 0, 0.52, 5]] /.
                Thread[{0, 1, 2} -> {RGBColor[0.6, 0.2, 0.1],
                                     RGBColor[0.1, 0.75, 0.2],
                                     RGBColor[1, 1, 0]}]]],
        AspectRatio -> Automatic];
```

When $f = 0$, forests spread as g increases and eventually there is a zero fire density and a percolation-like phase transition takes place at a critical value of g, which depends on p.

Identifying Forests and Measuring Forest Size Distribution

The size distribution of the clusters of trees (i.e., the forests) in the forest preserve is of interest. To determine this quantity, we first need to identify the various forests in the preserve.

The labeling procedure for the two-dimensional lattice having periodic boundaries can be explained using a simple Boolean lattice as an example.

```
(lat = {{1, 0, 1, 0, 0}, {0, 1, 0, 1, 0},
        {1, 1, 0, 1, 0}, {1, 0, 0, 0, 1}}) // MatrixForm
```

```
1   0   1   0   0
0   1   0   1   0
1   1   0   1   0
1   0   0   0   1
```

(i) The lattice sites at the northwest (i.e., top left) corners of the clusters are first labeled using

```
clusterID[lat, RotateRight[lat, {1, 0}], RotateRight[lat, {0, 1}]]
```

where

```
clusterCornerID[1, 0, 0] := i++
clusterCornerID[a_ , __] := a
Attributes[clusterCornerID] = Listable
```

The three arguments of `clusterCornerID` are the value of a site, the value of the nearest neighbor above the site, and the value of the nearest neighbor to the left of the site.

Using `clusterCornerID` with `lat`, we get

```
i = 2;
clusterCornerID[1, 0, 0]:= i++;
clusterCornerID[a_, __]:= a;
Attributes[clusterCornerID] = Listable;
(cornerLabels =
    clusterCornerID[#, RotateRight[#, {1, 0}],
                    RotateRight[#, {0, 1}]]&[lat]) // MatrixForm
```

```
1   0   2   0   0
0   3   0   4   0
5   1   0   1   0
1   0   0   0   6
```

Note: In performing the above labeling, the initial value of i is taken to be 2 (and as a result, cluster corner sites were numbered 2, 3, ...) because the value 1 is used to identify noncorner cluster sites.

(ii) Having labeled the sites at the corner of clusters, we next label the sites (those with value 1) that lie within clusters and we merge contiguous clusters. Both of these are accomplished with the following rules

```
reLabel[0, ___] := 0
reLabel[a_, b_, c_, d_, e_] := Max[a, b, c, d, e]
```

where the five arguments of reLabel are the value of a site, the value of the nearest neighbor above the site, the value of the nearest neighbor to the left of the site, the value of the nearest neighbor beneath the site, and the value of the nearest neighbor to the right of the site.

Using FixedPoint to apply reLabel repeatedly to cornerLabels until the cluster labels no longer change, we get

```
reLabel[0, ___] := 0;
reLabel[a_, b_, c_, d_, e_] := Max[a, b, c, d, e];
VonNeumann[func__, lat_] :=
    MapThread[func, Map[RotateRight[lat, #]&,
              {{0, 0}, {1, 0}, {0, -1}, {-1, 0}, {0, 1}}], 2];

forestLabels =
    FixedPoint[VonNeumann[reLabel, #]&, cornerLabels] // MatrixForm
```

```
6   0   2   0   0
0   6   0   4   0
6   6   0   4   0
6   0   0   0   6
```

Lastly, we can renumber the forests so that there are no gaps in the numbering, using

```
Thread[# -> Range[Length[#]]]&[Rest[Union[Flatten[forestLabels]]]]
```

```
{2 -> 1, 4 -> 2, 6 -> 3}
```

We can combine these code fragments into a labeling program.

```
clusterLabel[lat_List] :=
 Module[{i=2, clusterCornerID, cornerLabels, reLabel},

  clusterCornerID[1, 0, 0] := i++;
  clusterCornerID[a_, __] := a;
  Attributes[clusterCornerID] = Listable;

  cornerLabels = clusterCornerID[#, RotateRight[#, {1, 0}],
                               RotateRight[#, {0, 1}]]&[lat];
```

```
reLabel[0, ___] := 0;
reLabel[a_, b_, c_, d_, e_] := Max[a, b, c, d, e];
Attributes[reLabel] = Listable;

forestLabels =
  FixedPoint[reLabel[#, RotateRight[#, {1, 0}],
               RotateRight[#, {0, 1}], RotateRight[#,{-1, 0}],
               RotateRight[#, {0, -1}]]&,
           cornerLabels];

forestLabels /.
  Thread[# -> Range[Length[#]]]&
            [Rest[Union[Flatten[forestLabels]]]]
 ]
```

Note: In *Mathematica*, the clusterLabel program is a more efficient (and elegant) algorithm for identifying clusters of connected lattice sites than the classical Hoshen-Kopelman algorithm (which is described in section 5.2 in Gaylord and Wellin, 1995).

To identify the forests (which contains both trees and burning trees) using the clusterLabel program, we apply clusterLabel to the lattice configuration produced by the contagion program, after first changing the 2's to 1's.

```
forestID = clusterLabel[forestFire[n, s, k, p, f, g, t] /. 2 -> 1]
```

Note: Clusters of burning trees can be identified by replacing the transformation rule with $\{1 \to 0, 2 \to 1\}$.

Having identified the various forests, it is straightforward to calculate the size distribution of the forests. We first calculate the size of each tree cluster in the forest preserve using

```
forestSizes =
    Function[y, Map[Count[Flatten[y], #]&, Range[Max[y]]]][forestID]
```

and we then determine the frequency with which these sizes occur using

```
Function[x, Map[{Count[x, #], #}&, Union[x]]][forestSizes]
```

Computer Simulation Projects

1. Create a graphic of the spread of disease using the contagion program and the following coloring scheme: susceptible individuals (sites with

value 0) are blue, immune individuals (sites with value between $a + 1$ and g) are green, and infectious individuals (sites with value between 1 and a) are red, with the level of redness decreasing as the site value increases.

Note: This coloring scheme depicts infectious individuals as becoming less infectious with time.

2. In a stochastic version of the epidemic CA (Schonfisch, 1995), a susceptible individual becomes infectious with a probability proportional to the fraction of infectious nearest neighbors. Create a program for this stochastic epidemic CA.

    Note: This is a *contact process* model, which is known in high-energy physics as the reggeon spin model in reggeon field theory.

3. Modify the stochastic epidemic CA developed in the previous exercise so that a susceptible individual becomes infectious with a probability proportional to the total level of infectiousness among its nearest neighbors.

    Hint: assume that the level of infectiousness of an individual decreases with time so that an individual who has just become infectious (i.e., a site with value 1) is most infectious while an individual who has been infectious for a time steps (ie., a site with value a) is least infectious.

4. The most well-known excitable medium CA model of infectious spreading is the *hodgepodge machine*, consisting of an s-by-s square lattice with periodic boundary conditions, where the lattice sites, which are called cells, have values ranging from 0 to r. Cells having a value 0 are said to be healthy, cells having a value r are said to be ill, and all other cells (i.e., cells having a nonzero value less than the maximum value) are said to be infected (the higher the value, the more infected the cell is). Cells in the hodgepodge CA are updated based on their von Neumann neighborhoods, using the following rules:

    •   An ill cell becomes healthy.

        ```
        sick[r, _, _, _, _]   := 0
        ```

    •   A healthy cell becomes infected to a degree given by the integer sum of ((the number of ill cells in the neighborhood/k_1) + (the number of infected cells in the neighborhood/k_2)), where k_1 and k_2 are weighting factors that specify the "resistance" of cells to infection.

```
sick[0, b_, c_, d_, e_] :=
   Min[r, Floor[(Floor[N[b/r]] + Floor[N[c/r]] +
                 Floor[N[d/r]] + Floor[N[e/r]])/k1] +
            Floor[(Sign[Mod[b, r]] + Sign[Mod[c, r]] +
                   Sign[Mod[d, r]] + Sign[Mod[e, r]])/k2]
      ]
```

- An infected cell becomes more infected to a degree given by the integer sum of (g + (the sum of the values of the cells in the neighborhood/the number of infected cells in the neighborhood)), where g is an integer indicating the virulence or speed of infection.

```
sick[a_, b_, c_, d_, e_] :=
 Min[r, g + Floor[(a + b + c + d + e)/
          (Sign[Mod[a, r]] + Sign[Mod[b, r]] + Sign[Mod[c, r]]
          + Sign[Mod[d, r]] + Sign[Mod[e, r]])]]
    ]
```

Note: In one variation of the hodgepodge model, the sum of the values of all of the cells in the neighborhood, $(a + b + c + d + e)$, is replaced by the sum of the values of the infected cells in the neighborhood, $(\mathrm{Mod}[a, r] + \mathrm{Mod}[b, r] + \mathrm{Mod}[c, r] + \mathrm{Mod}[d, r] + \mathrm{Mod}[e, r])$.

The overall hodgepodge machine CA program is given by:

```
hodgepodge[r_Integer, s_Integer, k1_, k2_, g_Integer, t_Integer] :=
Module[{initconfig, VonNeumann, sick},

  initconfig = Table[Random[Integer, {0, r}], {s}, {s}];

  VonNeumann[func__, lat_] :=
    MapThread[func, Map[RotateRight[lat, #]&,
              {{0, 0}, {1, 0}, {0, -1}, {-1, 0}, {0, 1}}], 2];

  sick[r, _, _, _, _] := 0;

  sick[0, b_, c_, d_, e_] :=
    Min[r, Floor[(Floor[N[b/r]] + Floor[N[c/r]]  +
                  Floor[N[d/r]]  + Floor[N[e/r]])/k1] +
             Floor[(Sign[Mod[b, r]] + Sign[Mod[c, r]] +
                    Sign[Mod[d, r]] + Sign[Mod[e, r]])/k2]
        ];
```

```
sick[a_, b_, c_, d_, e_] :=
  Min[r, g + Floor[(a + b + c + d + e)/
            (Sign[Mod[a, r]] + Sign[Mod[b, r]] + Sign[Mod[c, r]]
            + Sign[Mod[d, r]] + Sign[Mod[e, r]])]
  ];


FixedPoint[VonNeumann[sick, #]&, initconfig, t]
]
```

5. The contagion model can be extended by incorporating mobility into the program. Combine the interactive random walkers CA with the contagion CA to create the "typhoid Mary" CA model of contagious disease spreading among a mobile population.

6. The sandpile CA has been suggested as the prototype for SOC (self-organized critical) behavior (Bak, 1991). The sandpile CA uses a two-dimensional square lattice with absorbing boundary conditions. Internal lattice sites have random values between 1 and 8 while border sites have a value of 0. Sites that have a value of 5 or more, are said to be "top-heavy." Starting with no top-heavy sites in the system, the values of randomly chosen sites are incremented by 1 until one of the sites reaches the "threshold" value of 5 and becomes top-heavy. At this point, the values of all of the sites are simultaneously updated by reducing the value of each top-heavy site by 4, and increasing the values of its four nearest-neighbor sites by 1. This is done repeatedly until there are no top-heavy sites left in the system.

The update rules for carrying out the "toppling" process are

- The value of a top-heavy site is decreased by 4 and increased by the number of nearest-neighbor sites that are top-heavy.

```
topple[a_, b_, c_, d_, e_] :=
      a - 4 + Floor[b/5] + Floor[c/5] + Floor[d/5] + Floor[e/5]
```

- The value of any other interior site is increased by the number of top-heavy nearest neighbor sites.

```
topple[a_ /; a < 5 , b_, c_, d_, e_] :=
      a + Floor[b/5] + Floor[c/5] + Floor[d/5] + Floor[e/5]
```

- The value of a border site remains 0.

```
topple[0, _, _, _, _ ] := 0
```

The overall sandpile CA program is given by

```
catastrophe[s_, m_]:=
 Module[{absorbBC, landscape, topple},

   absorbBC = (Prepend[Append[Map[Prepend[Append[#, 0], 0]&, #],
                              Table[0, {Length[#] + 2}]],
                      Table[0, {Length[#] + 2}]])&;

   landscape = absorbBC[Table[Random[Integer, {1, 4}], {s}, {s}]];

   While[Max[landscape]  <  5,
         randx = Random[Integer, {2, s + 1}, {2, s + 1}];
         randy = Random[Integer, {2, s + 1}, {2, s + 1}];
         landscape[[randx, randy]]++
         ];

   topple[0, _, _, _, _ ] := 0;
   topple[a_ /; a < 5 , b_, c_, d_, e_] :=
               a + Floor[b/5] + Floor[c/5] + Floor[d/5] + Floor[e/5];
   topple[a_, b_, c_, d_, e_] :=
               a - 4 + Floor[b/5] + Floor[c/5] + Floor[d/5] + Floor[e/5];

   VonNeumann[func__, lat_] :=
      MapThread[func, Map[RotateRight[lat, #]&,
               {{0, 0}, {1, 0}, {0, -1}, {-1, 0}, {0, 1}}], 2];

   FixedPointList[VonNeumann[topple, #]&, landscape, m]
 ]
```

An entertaining animation of the avalanching behavior of the sand-pile CA can be made (Gaylord and Wellin, 1995) by loading the *Mathematica* package

```
Needs["Graphics'Graphics3D'"]
```

and then using

```
ShowCatastrophe[list_, opts___]:=
   Map[(BarChart3D[list[[#]], PlotRange -> {0,8},opts])&,
       Range[Length[list]]]
```

References

Bak, Per. 1991. "Catastrophes and Self-Organized Criticality." Computers in Physics, 5(July/Aug), 430–433. 1991. "Self-Organized Criticality." Scientific American, 264(1), 46–53.

Dossel, B. and Schwabl, F. 1994. "Formation of space-time structure in a forest-fire model." Physica A, 204, 212–229.

Durret, Rick. 1991. "Some new games for your computer." Nonlinear Science Today, 1(4), 1–7.

Gaylord, Richard J. and Wellin, Paul R. *Computer Simulations with Mathematica: Explorations in Complex Physical and Biological Systems.* TELOS/Springer-Verlag (1995).

Maddox, John. 1992. "Forest fires, sandpiles and the like." Nature, 359, 359.

Madore, B.F. and Freedman, W.L. 1987. "Self-organizing structures." American Scientist, 75, 252–259.

Schonfisch, Brigitt. 1995. "Propagation of fronts in cellular automata." Physica D, 80, 433–450.

16 The Evolution of Cooperation and the Spatial Prisoner's Dilemma Game

Introduction

Why do individuals, people and animals, help one another? Restated in the terminology of evolutionary biology, what is the survival value of altruism (i.e., why does natural selection result in the evolution of cooperative social behavior)?

The question of the origin of cooperative social behavior is not unimportant. Essentially every political philosophy, with the exception of radical libertarianism (anarcho-capitalism), is based on the axiom that it is necessary to have a powerful authority, the state, to bring about cooperation between individuals. If it can be shown that social cooperation arises naturally among people without being forced upon them, then the rationale often given for government intrusion in the affairs of its citizens becomes less compelling.

What makes the evolution of cooperation between members in a group especially puzzling is that whenever two individuals interact in a single encounter, the rational behavior for each party is to be non-cooperative. This can be shown by a game theoretic analysis of the situation known as the prisoner's dilemma. In an isolated interaction between two individuals, each of whom must choose between two interaction strategies, cooperation and non-cooperation, there are four possible outcomes of the interaction: a win-win situation arising from both parties (players) cooperating with one another (each player gets a payoff R), a win-lose situation and a lose-win situation in which one of the parties cooperates while the other party does not (the uncooperative player gets a payoff

T and the cooperative player gets a payoff S), and a lose-lose situation in which both parties are uncooperative (each player gets a payoff P). If $T > R > P > S$, it is always in each player's self-interest to be uncooperative.

Several solutions have been proposed to account for what appears to be irrational "nice guy" behavior between the unrelated individuals in a group (altruism between kinfolk is not unexpected; it can be understood in terms of gene survivability). Mutual aid behavior could be expected if there are repeated encounters between individuals, in each of which, the players choose their strategies based either on the outcomes obtained in previous encounters or in anticipation of future encounters. Cooperation might also arise if natural selection were concerned with the survival of groups rather than individuals.

Recently, it has been suggested (Nowak and May, 1992–5) that spatial effects alone are sufficient to cause cooperative behavior. Nowak and May use a cellular automaton to model the evolution of cooperation. In their CA, all of the sites of a two-dimensional lattice are occupied by players. The players interact with their nearest neighbor players in a pair-wise manner, over a number of time steps. The interaction strategies used by the players in each time step is determined as follows: In a given time step, each player interacts with itself and with its eight nearest neighbors (i.e., the nine sites in the Moore neighborhood of the player) using either a cooperative strategy or an uncooperative strategy. The total payoff to each player resulting from the nine interactions is determined, and each player then adapts for the next time step the strategy of the player in its Moore neighborhood (including itself) who received the biggest payoff.

We will develop a program for this "spatial prisoner's dilemma" cellular automaton.

The Spatial Prisoner's Dilemma CA

The strategy adopted in each time step by a player (i.e., the value of each site in the CA lattice), depends on both the total payoff earned by the player as a result of its pair-wise interactions with its eight nearest neighbor players and the total payoffs earned by the nearest neighbor players as a result of their interactions with their eight nearest neighbor players. Thus, each player's strategy in a given time step is based the values of the 25 neighbors in its 5-by-5 neighborhood. Since each player can have two values (representing cooperation or non-cooperation), a specification of the CA rules based on enumerating every possible set of values of the 25 neighbor sites would require 2^{25} rules. Rather attempting to write down or generate these (approximately) 33,000,000 rules for use in a lookup table in the CA program, we will employ a more computational approach to implementing the CA.

The System

The CA employs a $(2n+1)$-by-$(2n+1)$ square lattice, with periodic boundary conditions. The lattice sites have values of 1 or 0, where 1 represents a strategy of cooperation and 0 represents a strategy of non-cooperation (defection). The initial distribution of cooperators and defectors needs to be specified. For a random distribution of cooperators and defectors on the lattice, the initial configuration is given by

```
initConf = Table[Random[Integer], {2 n + 1}, {2 n + 1}]
```

For a single defector amid a sea of cooperators, initConf is given by

```
initConf =
  ReplacePart[Table[1, {2 n + 1}, {2 n + 1}], 0, {n + 1, n + 1}]
```

Playing the Game

The prisoner's dilemma game is played over a number of rounds (i.e., time steps), in each of which every prisoner (site) on the lattice interacts with itself and with the eight nearest neighbor sites in its Moore neighborhood.

Each round of interaction proceeds as follows (we will use the symbol jail below to indicate the value of the CA lattice in a given round of play):

In the first stage, each prisoner interacts with the prisoners in its Moore neighborhoods in a pair-wise manner. A *payoff* value is computed for each interaction as follows:

If the player and its neighbor both cooperate, the player gets a point: [cooperate − cooperate → 1]

If the player and its neighbor both defect, the player gets nothing: [defect − defect → 0]

If the player cooperates and its neighbor defects, the player gets nothing: [cooperate − defect → 0]

If the player defects and its neighbor cooperates, the player gets p points $(p > 1)$: [defect − cooperate → p]

Note: These payoff values correspond to using $S = P = 0$, $R = 1$, $T = p$ $(p > 1)$ in the prisoner's dilemma. These values are used to simplify the program to using a single payoff parameter value, but other payoff values will produce the same results (see computer simulation project #2).

The total payoff for a player in a round is the sum of the nine pay-offs resulting from the interactions of the player with each site in its neighborhood (including itself). Since 1 represents a cooperator and 0 represents a defector in the model, the total payoff to a cooperator site in jail is the number of cooperator sites in its Moore neighborhood and the total payoff to a defector site in jail is p times the number of cooperator sites in its Moore neighborhood.

The following `totalPayoff` rules whose arguments are the values of the nine players in the Moore neighborhood of a site are used:

```
totalPayoff[1, a_, b_, c_, d_, e_, f_, g_, h_] :=
                    {1 + a + b + c + d + e + f + g + h, 1}

totalPayoff[0, a_, b_, c_, d_, e_, f_, g_, h_] :=
                    {p (a + b + c + d + e + f + g + h), 0}
```

The totalPayoffPlayer rules are applied to the neighborhood of each player in jail, using the following function:

```
Moore[totalPayoff, #]&
```

where

```
Moore[func__, lat_] :=
  MapThread[func, Map[RotateRight[lat, #]&,
          {{0, 0}, {1, 0}, {0, -1}, {-1, 0}, {0, 1},
          {1, -1}, {-1, -1}, {-1, 1}, {1, 1}}], 2]
```

Applying this anonymous function to the jail produces a lattice whose site values are ordered pairs, where the second component is the strategy used by the corresponding player in jail, and the first component is the total payoff received by that player as a result of the pair-wise interactions with its Moore neighbors.

The lattice whose site values consist of the strategy used by the player in the neighborhood of the corresponding site in jail who receives the greatest payoff from its pair-wise neighbor interactions is determined by applying the following function to jail:

```
Moore[Last[Sort[{##}]][[2]]&, #]&[Moore[totalPayoff, #]]&
```

When this function is applied to jail, the following operations occur, in order:

The ordered pairs consisting of the total payoff to each player in jail and its strategy are created.

The nine ordered pairs for each Moore neighborhood in jail are placed in a list.

The elements of the list are reordered so that the components of the last ordered pair in the list are the highest payoff and the strategy used by the player who received that payoff.

The strategy of the player receiving the highest payoff is extracted from the last ordered pair.

The system evolves by applying this anonymous function repeatedly to jail *t* times using the NestList function

```
NestList[Moore[Last[Sort[{##}]][[2]]&, #]&[Moore[totalPayoff, #]]&,
        initConf, t]
```

The Program

We'll look at the evolution of a system consisting of a single defector in a sea of collaborators.

```
SpatialPrisonersDilemma[n_, p_, t_] :=
  Module[{initConf, Moore, totalPayoff},

  initConf =
    ReplacePart[Table[1, {2 n + 1}, {2 n + 1}], 0, {n + 1, n + 1}];

  Moore[func__, lat_] :=
    MapThread[func, Map[RotateRight[lat, #]&,
              {{0, 0}, {1, 0}, {0, -1}, {-1, 0}, {0, 1},
              {1, -1}, {-1, -1}, {-1, 1}, {1, 1}}], 2];

  totalPayoff[1, a_, b_, c_, d_, e_, f_, g_, h_] :=
                        {1 + a + b + c + d + e + f + g + h, 1};
  totalPayoff[0, a_, b_, c_, d_, e_, f_, g_, h_] :=
                        {p (a + b + c + d + e + f + g + h), 0};

  NestList[Moore[Last[Sort[{##}]][[2]]&, #]&[Moore[totalPayoff, #]]&,
          initConf, t]
  ]
```

Note: It is interesting to do a side-by-side comparision of programs that implement the spatial prisoner's dilemma model in a functional, rule-based style, as done here, and in a procedural style (Lloyd, 1995).

"Book 'em, Dano"

To visualize the results obtained from running the `SpatialPrisoners-Dilemma` program, we can create a `RasterArray` using a coloring scheme based on the strategy used by each player in both the current round and the previous round. The coloring is done as follows:

If a player is cooperating now and cooperated in the previous step, it is colored blue.

If a player is defecting now and defected in the previous step, it is colored red.

If a player is cooperating now and defected in the previous step, it is colored green.

If a player is defecting now and cooperated in the previous step, it is colored yellow.

According to this scheme, green and yellow cells indicate changes in strategy from one generation to the next, and red and blue indicate cells with unchanged strategies.

The coloring scheme is implemented in the following animation program:

```
SocialEvolution[lis_] :=
 Module[{picture},

      picture = Rest[MapThread[color, {#, RotateRight[#]}&[lis], 3]];

      Map[Show[Graphics[RasterArray[Reverse[#] /.
                          {color[1, 1] -> RGBColor[0, 0, 1],
                           color[0, 0] -> RGBColor[1, 0, 0],
                           color[1, 0] -> RGBColor[0, 1, 0],
                           color[0, 1] -> RGBColor[1, 1, 0]}]],
              AspectRatio -> Automatic, DisplayFunction -> Identity]&,
          picture]
   ]
```

A single graphic showing all of the animation cells created with `SocialEvolution` can be created using

```
Show[GraphicsArray[Partition[SocialEvolution[lis], r] ]]
```

An example output is shown below for the case of an initial state of a single defector in a sea of cooperators.

```
Show[GraphicsArray[Partition[
        SocialEvolution[SpatialPrisonersDilemma[25, 1.85,  20]], 2] ]];
```

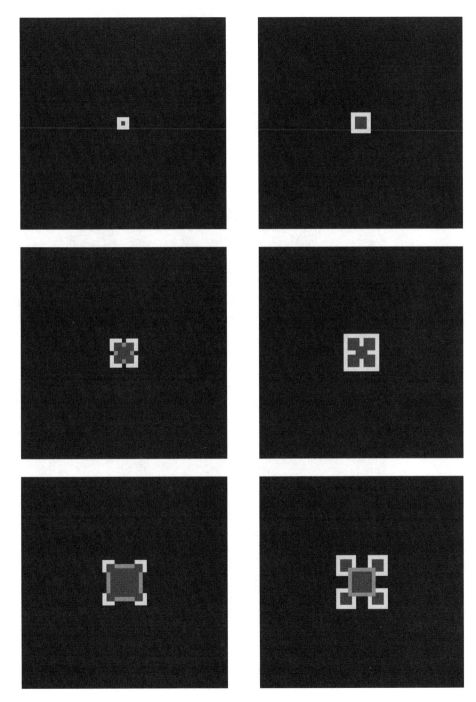

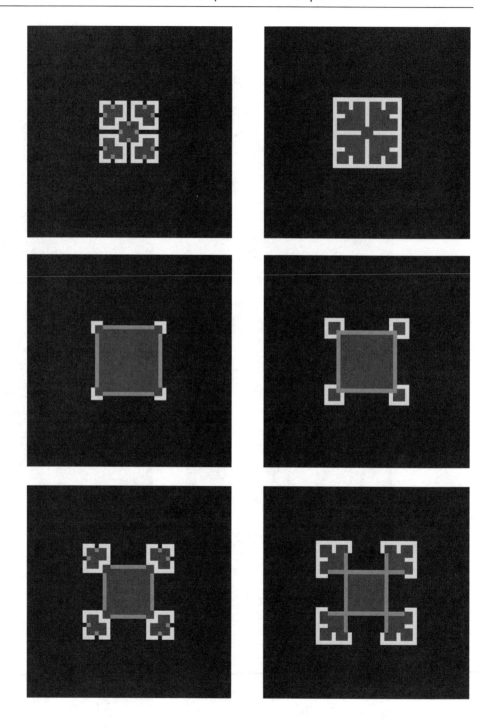

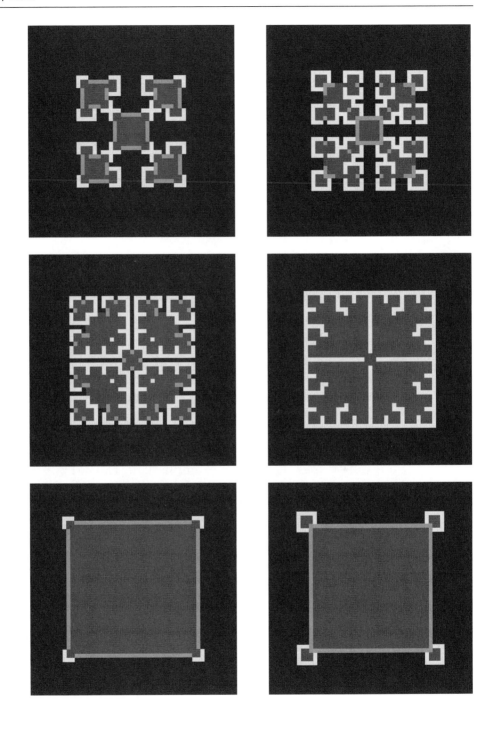

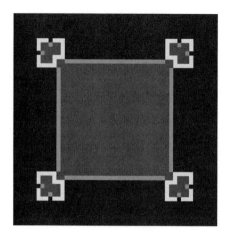 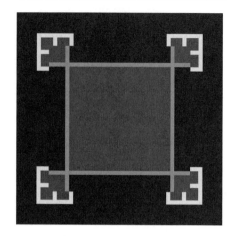

For an initial state consisting of a random distribution of cooperators or defectors, cooperators and defectors are found to coexist in fluctuating chaotic patterns for a range of b values.

Computer Simulation Projects

1. Calculate and plot the frequency of cooperators in the spatial prisoner's dilemma game over time as a function of p, the advantage of defection over cooperation.

2. In addition to "spatializing" the prisoner's dilemma, other evolutionary games can be given a spatial component by using various payoff values in the program. For example, in the Hawk-Dove game, the payoff for a hawk interacting with another hawk is less than the payoff for a dove interacting with a hawk while the payoff for a dove interacting with a dove is less than the payoff for a hawk interacting with a dove. Modify the outcome rules to implement the spatial Hawk-Dove game and run the resulting program.

3. There are two interaction strategies used in the prisoner's dilemma. A variation of the game is obtained by having three or more strategies. One example of this is the well-known children's game, "stone-scissors-paper," in which stone beats scissors, scissors beats paper, and paper beats stone. Implement this game and run it to obtain spiral waves, reminiscent of excitable media behavior (see the Contagion chapter).

4. An interesting way to extend the spatial evolutionary game would be to incorporate diffusive movement by the players. This can be done by combining the interacting random walk CA with the spatial prisoner's dilemma CA.

References

Lloyd, Alun, L. 1995. "Computing bouts of the prisoner's dilemma." Scientific American, 272, 110–115.

Nowak, Martin A., Bonhoeffer, Sebastian and May, Robert, M. 1994. "More spatial games." International Journal of Bifurcation and Chaos, 4, 33–56.

Nowak, Martin A. and May, Robert, M. 1992. "Evolutionary games and spatial chaos." Nature, 359, 826–829.

Nowak, Martin A. and May, Robert, M. 1993. "The spatial dilemma of evolution." International Journal of Bifurcation and Chaos, 3, 35–78.

Nowak, Martin A., May, Robert, M., and Sigmund, Karl. 1995. "The arithmetics of mutual help." Scientific American, 272, 76–81.

A Mathematica Programming Tutorial

Introduction

To use *Mathematica* efficiently, you need to understand how the *Mathematica* programming language works. This tutorial (which has been extensively field-tested in university courses and in one-day *Mathematica* programming courses) is intended to provide you with the background that's needed to understand the code in this book and to write your own code for cellular automata explorations.

Note: While this material can't replace the use of the *Mathematica* book by Wolfram or *Mathematica*'s on-line help (which gives definitions of built-in functions and simple illustrations of their use), it can help to make you more comfortable with the *Mathematica* programming language, which often seems obscure, even engimatic, when first encountered by someone whose programming experience is with one of the traditional procedural languages.

In this note set, the following aspects of the *Mathematica* programming language are emphasized: the nature of expressions, how expressions are evaluated, how pattern-matching works, creating rewrite rules, and using higher-order functions.

Summing the Elements in a List

Consider the data structure {1, 2, 3}. How can we add up the elements in the list?

```
Apply[Plus, {1, 2, 3}]
```

```
6
```

What's going on here?

Everything is an Expression

Every quantity entered into *Mathematica* is represented internally as an expression.

An expression has the form

$$\text{head[arg1, arg2, ..., argn]}$$

where the "head" and "arguments" can be other expressions.

For example, if we look at two common quantities, a list data structure $\{a, b, c\}$, and an arithmetic operation $a + b + c$, they appear to be quite different, but if we use the FullForm function to look at their internal representation

```
FullForm[{a, b, c}]
```

```
List[a, b, c]
```

```
FullForm[a + b + c]
```

```
Plus[a, b, c]
```

we see that they differ only in their heads.

The use of a common expression structure to represent everything is not merely cosmetic; it allows us to perform some computations quite simply. For example, to add the elements in a list, it is only necessary to change the head of the expression, List, to Plus. This can be done using the built-in Apply function.

```
?Apply
```

```
Apply[f, expr] or f @@ expr replaces the head of expr by f.
Apply[f, expr, levelspec] replaces heads in parts of expr specified by
levelspec.
```

```
Trace[Apply[Plus, {1, 2, 3}]]
```

```
{Apply[Plus, {1, 2, 3}], 1 + 2 + 3, 6}
```

Changing a Sum of Elements into a List of Elements

The obvious approach to this task is to do the same sort of thing that we did to add the elements in a list.

```
Apply[List, a + b + c]
```

```
{a, b, c}
```

This works when the list elements are symbols, but it doesn't work for a list of numbers:

```
Apply[List, 1 + 2 + 3]
```

```
6
```

To understand the reason for the different results obtained above, it is necessary to understand how *Mathematica* evaluates expressions. To do that, we first want to distinguish between the various kinds of expressions.

Kinds of Expressions

Non-Atomic Expressions

Non-atomic expressions have parts that can be extracted from the expression with the Part function, and can be replaced with the ReplacePart function. For example,

```
Part[{a, 7, c}, 1]
```

```
a
```

```
{a, 7, c}[[0]]
```

```
List
```

```
Part[a + b + c, 0]
```

```
Plus
```

```
ReplacePart[{a, 7, c}, e, 2]
```

```
{a, e, c}
```

Atomic Expressions

Atomic expressions constitute the basic building blocks of the *Mathematica* language.

There are three kinds of atomic expressions:

(1) A symbol, consisting of a letter followed by letters and numbers (e.g., darwin)

(2) Four kinds of numbers:

integer numbers (e.g., 4)
real numbers (e.g., 5.201)
complex numbers. (e.g., 3 + 4*I*)
rational numbers (e.g., 5/7)

(3) A string comprised of letters, numbers, and spaces (i.e., ASCII characters) between quotes (e.g., "Computer Simulations with Mathematica")

Atomic expressions differ from non-atomic expressions in several ways:

The `FullForm` of an atomic expression is the atom itself.

```
{FullForm[darwin], FullForm["Computer Simulations with Mathematica"],
 FullForm[5]}
```

```
{darwin, "Computer Simulations with\
   Mathematica", 5}
```

The head (or 0th part) of an atom is the type of atom that it is.

```
{Head[List], Head["Computer Simulations with Mathematica"], 5[[0]]}
```

```
{Symbol, String, Integer}
```

An atomic expression has no parts that can be extracted or replaced.

```
Part["Computer Simulations with Mathematica", 1]
```

```
Part::partd:
  Part specification
   Comput<<28>>ica[[1]]
     is longer than depth of object.
```

Compound Expressions

A CompoundExpression is an expression consisting of a succession of expressions separated by semicolons (;).

```
expr1; expr2; ...; exprn
```

```
5 + 3; 7 4
```

Entering an Expression

When an expression is entered in *Mathematica*, it is evaluated and the result is returned, unless it is followed by a semicolon.

```
4^3
```

```
64
```

When an expression is followed by a semicolon, the expression is also evaluated, even though nothing is returned.

```
2 - 6;
```

```
% + 3
```

```
-1
```

```
%%
```

```
-4
```

When the entered expression is a compound expression, its components are evaluated sequentially and the result of the last evaluation is returned.

```
Trace[a = 3 + 5; 4 a]
```

```
{a = 3 + 5; 4 a, {{3 + 5, 8}, a = 8, 8},
   {{a, 8}, 4 8, 32}, 32}
```

How Expressions are Evaluated

Mathematica is a term rewriting system (TRS). Whenever an expression is entered, it is evaluated by term rewriting using rewrite rules. These rules consist of two parts: a pattern on the left-hand side (lhs) and a replacement text on the right-hand side (rhs). When the lhs of a rewrite rule is found to pattern-match part of the expression, that part is replaced by the rhs of the rule, after substituting values in the expression that match labeled blanks in the pattern into the rhs of the rule. Evaluation then proceeds by searching for further matching rules until no more are found.

The evaluation process in *Mathematica* can be easily understood with the following analogy:

Think of your experiences with using a handbook of mathematical formulas.

To solve an integral, you consult the handbook, which contains formulas consisting of a left-hand side (lhs) and a right-hand side (rhs), separated by an equal sign.

You look for an integration formula in the handbook whose left-hand side has the same form as your integral.

Note: While no two formulas in the handbook have the identical lhs, there may be several whose lhs have the same form as your integral (e.g., one lhs might have specific values in the integration limits or in the integrand, while another lhs has unspecified (dummy) variables for these quantities). When this happens, you use the formula whose lhs gives the closest fit to your integral.

You then replace your integral with the rhs of the matching lhs and you substitute the specific values in your integral for the corresponding variable symbols in the rhs.

Finally, you look through the handbook for formulas (e.g., trigonometric identities or algebraic manipulation) that can be used to change the answer further.

This depiction provides an excellent description of the *Mathematica* evaluation process.

However, the application of the term-rewriting process to an *Mathematica* expression requires a bit more discussion because a *Mathematica* expression consists of parts, a head, and zero or more arguments that are themselves expressions.

```
expr0[expr1, expr2, ..., exprn]
```

It is therefore necessary to understand the order in which the various parts of an expression are evaluated by term rewriting.

The implementation of the evaluation procedure is (with a few exceptions) straightforward:

1. If the expression is a number or a string, it isn't changed.
2. If the expression is a symbol, it is rewritten if there is an applicable rewrite rule in the global rule base; otherwise, it is unchanged.
3. If the expression is not a number, string, or symbol, its parts are evaluated in a specific order.

   a. The head of the expression is evaluated.
   b. The arguments of the expression are evaluated from left to right in order. An exception to this occurs when the head is a symbol with a Hold attribute (e.g., HoldFirst, HoldRest, or HoldAll), so that some of its arguments are left in their unevaluated forms (unless they, in turn, have the head Evaluate).

4. After the head and arguments of an expression are each completely evaluated, the expression consisting of the evaluated head and arguments is rewritten, after making any necessary changes to the arguments based on the Attributes (such as Flat, Listable, Orderless) of the head, if there is an applicable rewrite rule in the global rule base.
5. After carrying out the previous steps, the resulting expression is evaluated in the same way and then the result of that evaluation is evaluated, and so on until there are no more applicable rewrite rules.

The details of the term-rewriting process in steps 2 and 4 are as follows:

a. part of an expression is pattern-matched by the lhs of a rewrite rule.
b. the values that match labeled blanks in the pattern are substituted into the rhs of the rewrite rule and evaluated.
c. the pattern-matched part of the expression is replaced with the evaluated result.

With this understanding of the evaluation procedure, we can now understand what happened when we entered

```
Apply[List, 1 + 2 + 3]
```

6

In evaluating this expression, the argument $1 + 2 + 3$ was evaluated before the Apply function was employed.

```
Trace[Apply[List, 1 + 2 + 3]]
```

```
{{1 + 2 + 3, 6}, Apply[List, 6], 6}
```

Controlling the Evaluation Order

We should point out that the user can (to some extent) wrest control of the evaluation process from *Mathematica* and either force or prevent evaluation. We won't go into the details of doing this but we can indicate functions that can be used for this purpose: Hold, HoldAll, HoldFirst, HoldRest, HoldForm, HeldPart, ReleaseHold, Evaluate, Unevaluated, and Literal.

To turn the sum into a list, it is necessary to prevent the argument Plus[1, 2, 3] from being prematurely evaluated before the symbol Plus is replaced with the symbol List.

```
Apply[List, Unevaluated[1 + 2 + 3]]
```

```
{1, 2, 3}
```

Since term rewriting is based on pattern-matching, we need to look at the various sorts of patterns that *Mathematica* recognizes.

Patterns

Blanks

Patterns are defined syntactically, i.e., by the internal representation of an expression as given using FullForm.

In general, an expression will be matched by several patterns of differing specificity. For example, constructing as many patterns to match x^2, in order of increasing generality

1. *x* raised to the power of two.
2. *x* raised to the power of a number.
3. *x* raised to the power of something.
4. a symbol raised to the power of two.
5. a symbol raised to the power of a number.
6. a symbol raised to the power of something.
7. something raised to the power of two.
8. something raised to the power of a number.

9. something raised to the power of something.

10. something.

The term *something* used above can be replaced by the term *an expression*, so that for example, the last case says that x^2 pattern-matches an expression (which is true since x^2 is an expression). To be precise, we need a notation to designate a pattern that has the form of an expression. We also need to designate a pattern that has the form of a sequence of expressions, consecutive expressions separated by commas.

Patterns are defined in *Mathematica* as expressions that may contain blanks. A pattern may contain a single (_) blank, a double (__) blank, or a triple (___) blank (the differences will be discussed shortly).

Note: A pattern can be labeled (given a name) by preceding the blanks(s) by a symbol, e.g., name_ or name__ or name___. The labeled pattern is matched by exactly the same expression that matches its unlabeled counterpart (pattern labeling, as we will see, is used to create dummy variables).

Note: A blank can be followed by a symbol, e.g., _h or __h or ___h, in which case an expression must have the head *h* to match the pattern (this is used to perform type checking).

Pattern-Matching an Expression

We can use the `MatchQ` function to determine if a particular pattern matches an expression or a sequence of expressions. The most specific pattern-match is between an expression and itself.

```
MatchQ[x^2, x^2]
```

```
True
```

To make more general (less specific) pattern-matches, a single blank is used to represent an individual expression, which can be any data object.

We'll work with x^2 to demonstrate the use of the `Blank` function in pattern-matching. In the following examples (which are arbitrarily chosen from the many possible pattern matches), we'll first state the pattern-match and then check it using `MatchQ`.

x^2 pattern-matches "an expression."

```
MatchQ[x^2, _]
```

```
True
```

x^2 pattern-matches "x raised to the power of an expression."

```
MatchQ[x^2, x^_]
```

True

x^2 pattern-matches "x raised to the power of an integer" (to put it more formally, "x raised to the power of an expression whose head is Integer").

```
MatchQ[x^2, x^_Integer]
```

True

x^2 pattern-matches "an expression whose head is Power":

```
MatchQ[x^2, _Power]
```

True

x^2 pattern-matches "an expression whose head is a symbol and that is raised to the power 2".

```
MatchQ[x^2, _Symbol^2]
```

True

x^2 pattern-matches "an expression raised to the power 2."

```
MatchQ[x^2, _^2]
```

True

x^2 pattern-matches "an expression whose head is a symbol and which is raised to the power of an expression whose head is an integer" (or stated less formally, "a symbol raised to the power of an integer").

```
MatchQ[x^2, _Symbol^_Integer]
```

True

x^2 pattern-matches "an expression raised to the power of an expression."

```
MatchQ[x^2, _^_]
```

```
True
```

x^2 pattern-matches "x raised to the power of an expression" (the label y does not affect the pattern-match).

```
MatchQ[x^2, x^y_]
```

```
True
```

Pattern-Matching a Sequence of One or More Expressions

A sequence consists of a number of expressions separated by commas. A double blank represents a sequence of one or more expressions and __h represents a sequence of one or more expressions, each of which has head h.

For example, a sequence in a list pattern-matches a double blank (note: we are pattern-matching the sequence in the list, not the list itself):

```
MatchQ[{a, b, c}, {__}]
```

```
True
```

but the arguments of an empty list (which has no arguments) do not pattern-match a double blank:

```
MatchQ[{}, {__}]
```

```
False
```

An expression that pattern-matches a blank also pattern-matches a double blank. For example,

```
MatchQ[x^2, __]
```

```
True
```

Pattern-Matching a Sequence of Zero or More Expressions

A triple blank represents a sequence of zero or more expressions and ___h represents a sequence of zero or more expressions, each of which has the head h. For example, the triple blank pattern-matches the empty list.

```
MatchQ[{}, {___}]
```

```
True
```

An expression that pattern-matches a blank and a sequence that pattern-matches a double blank pattern both pattern-match a triple blank pattern.

```
MatchQ[x^2, ___]
```

```
True
```

It is important to be aware that for the purposes of pattern-matching, a sequence is not an expression. For example,

```
MatchQ[{a, b, c}, {_}]
```

```
False
```

Alternative Pattern-Matching

We can make a pattern-match less restrictive by specifying alternative patterns that can be matched.

```
MatchQ[x^2, {_} | _^2]
```

```
True
```

Conditional Pattern-Matching

We can make a pattern-match more restrictive by making it contingent upon meeting certain conditions (note: satisfying these conditions will be a necessary, but not sufficient, requirement for a successful pattern-match).

If the blanks of a pattern are followed by ?test, where test is a predicate (i.e., a function that returns a True or False), then a pattern-match is only possible if test returns True when applied to the entire expression. ?test is used with built-in predicate functions and with anonymous predicate functions.

```
MatchQ[x^2, _^_?OddQ]
```

False

```
MatchQ[2, _?(# > 3&)]
```

False

```
MatchQ[2, _?(# > 1.5 &)]
```

True

```
MatchQ[2, _Integer?(# > 3&)]
```

False

If part of a labeled pattern is followed by /; *condition*, where *condition* contains labels appearing in the pattern, then a pattern-match is possible only if condition returns True when applied to the **labeled parts** of an expression. For example,

```
MatchQ[x^2,_^y_]
```

True

```
MatchQ[a^b,_^y_/; Head[y] == Symbol]
```

True

With this understanding of how pattern-matching works in *Mathematica*, we can discuss how to create our own rewrite rules which can be used in term rewriting.

Rewrite Rules

Built-in Functions

Mathematica provides nearly one thousand built-in functions that can be used for term rewriting. These rules are located in the global rule base whenever *Mathematica* is running. Functions defined in a *Mathematica* package are also placed in the global rule base during the session in which the package is loaded. Functions in the global rule base are always available for term rewriting and they are always used whenever applicable.

User-Defined Functions

In addition to the built-in rewrite rules, user-defined rewrite rules can be created and placed in the global rule base where they are always available, and always used, when applicable for the duration of the ongoing session. However, they are not automatically preserved beyond the session in which they are created.

There are basically two ways to create a user-defined rewrite rule: with the Set function and with the SetDelayed function.

Global Rules

Declaring a Value Using the Set (=) Function

A value declaration is essentially a nickname for a value (e.g., for a list or number) that can be used in place of the value. It is written using Set[lhs, rhs] or, more commonly, as

$$lhs = rhs$$

The lhs starts with a name, starting with a letter followed by letters and/or numbers (with no spaces). The rhs is either an expression or a compound expression enclosed in parentheses.

Note: the name on the lhs may be followed by a set of square brackets containing a sequence of patterns or labeled patterns, and the rhs may contain the labels, without the blanks.

For example, consider the following two simple Set functions

```
a = {-1, 1}

{-1, 1}

rand1 = Random[Integer, {1, 2}]

2
```

Notice that when a Set function is entered, a value is returned (unless it is followed by a semicolon). If we look into the global rule base to see what rewrite rules have been created when *a* and *rand1* were entered

```
?a
```

```
Global'a
```

```
a = {-1, 1}
```

```
?rand1
```

```
Global'rand1
```

```
rand1 = 2
```

We find that the rewrite rule associated with *a* is the same as the Set function we entered, but the rewrite rule associated with *rand1* differs from the corresponding Set function. The reason for this is that when a Set function is entered, its lhs is left unevaluated while its rhs is evaluated. This property is known as the HoldFirst attribute.

```
Attributes[Set]
```

```
{HoldFirst, Protected}
```

```
?HoldFirst
```

```
HoldFirst is an attribute which specifies that the first argument
to a function is to be maintained in an unevaluated form.
```

When a rewrite rule is created from a Set function, the unevaluated lhs and the evaluated rhs of the function are used.

In the above cases, the evaluation of the rhs of *a* resulted in $\{-1, 1\}$ and the evaluation of the rhs of *rand1* resulted in 2.

When the rhs is a compound expression enclosed in parentheses, the expressions of the rhs are evaluated in sequence and the rhs of the resulting rewrite rule is the result of the final evaluation. For example,

```
rand2 = (b = {-1, 1}; Random[Real, b])
```

```
-0.117994
```

```
?rand2
```

```
Global'rand2
```

```
rand2 = -0.1179941207497524877
```

What happened here is that the b was first evaluated to give $\{-1, 1\}$ and this value was then used to evaluate the random number function.

The order of expressions on the rhs is important. An expression on the rhs must appear before it is used in another expression on the rhs. For example,

```
rand3 = (Random[Real, c]; c = {-1, 1})
```

```
Random::randn:
   Range specification c in
    Random[Real, c]
     is not a valid number or pair of
     numbers.
```

```
{-1, 1}
```

When a Set function is entered, both it and any Set or SetDelayed functions on the rhs create rewrite rules in the global rule base.

```
?b
```

```
Global'b
```

```
b = {-1, 1}
```

```
?c
```

```
Global'c
```

```
c = {-1, 1}
```

After a value has been declared by entering a Set function, the appearance of the value's name during an evaluation causes the value itself to be substituted in (which is why we say that it acts like a nickname). For example,

```
Abs[rand2]
```

```
0.117994
```

What happened here was that the rewrite rule associated with *rand2* in the global rule base was used as an argument to the Abs function.

The lhs of a rewrite rule can only be associated with one value at a time. When a Set function is entered, the resulting rewrite rule "overwrites" any previous rewrite rule with the identical lhs. For example,

```
rand4 = Random[Integer, {1, 2}];
```

```
?rand4
```

```
Global'rand4
```

```
rand4 = 2
```

```
rand4 = Random[Integer, {1, 2}];
```

```
?rand4
```

```
Global'rand4
```

```
rand4 = 1
```

What we see is that the value of *rand4* was 2 after *rand4* was first entered and this value was then changed to 1 after *rand4* was reentered.

While the lhs of a rewrite rule can only be associated with one value at a time, a value can be associated with several names, simultaneously. We made use of this earlier when we defined both *b* and *c* as $\{-1, 1\}$.

Finally, user-defined rewrite rules can be removed from the global rule base using either the Clear or Remove function.

```
Clear[b]
```

```
?b
```

```
Global'b
```

```
Remove[c]
```

```
?c
```

```
Information::notfound:
    Symbol c not found.
```

Defining a Function Using the SetDelayed (:=) Function

Function definitions (i.e., programs) are written as

```
name[arg1_, arg2_, ..., argn_] := (expr1; expr2 ; ...; exprm)
```

The lhs starts with a name. The name is followed by a set of square brackets containing a sequence of labeled patterns, which are symbols ending with one or more underspaces (i.e., blanks). The rhs is either an expression or a compound expression enclosed in parentheses, containing the labels on the lhs (without the blanks).

For example, we'll enter the function definition

```
f[x_] := Random[Real, {0, x}]
```

The first thing we notice is that, in contrast to a Set function, nothing is returned when a SetDelayed function is entered. If we query the rule base,

```
?f
```

```
Global`f
```

```
f[x_] := Random[Real, {0, x}]
```

we see that a rewrite rule associated with *f* has been placed in the global rule base that is identical to the SetDelayed function. The reason is that when a SetDelayed function is entered both its lhs and rhs are left unevaluated. This property is known as the HoldAll attribute.

```
Attributes[SetDelayed]
```

```
{HoldAll, Protected}
```

```
?HoldAll
```

```
HoldAll is an attribute that specifies that all arguments to a
function are to be maintained in an unevaluated form.
```

A user-defined function is called in the same way as a built-in function is called, by entering its name with specific argument value(s).

```
f[8]
```

```
6.82692
```

Each time the lhs of a `SetDelayed` rewrite rule is entered with specific argument values, the rhs of the rule is evaluated using these values, and the result is returned.

Note that the rhs is evaluated afresh each time the lhs is entered with specific argument values. Thus,

```
f[8]
```

```
6.17498
```

When the rhs of the `SetDelayed` function is a compound expression enclosed in parentheses, no rewrite rules are created from the auxillary functions on the rhs when the function is entered (this is because the rhs is not evaluated).

```
g[x_] := (d = 2; x + d)
```

```
?g
```

```
Global'g
```

```
g[x_] := (d = 2; x + d)
```

```
?d
```

```
Global'd
```

When the program is run (or equivalently, a user-defined function is called) for the first time, all of the auxillary functions are then placed in the global rule base.

```
g[3]
```

```
5
```

```
?d
```

```
Global'd
```

```
d = 2
```

Placing Constraints on a Rewrite Rule

The use of a rewrite rule can be restricted by attaching constraints on either the lhs or the rhs of a `SetDelayed` rule. Conditional pattern-matching with _h or with _? and _/; can be attached to the dummy variable arguments on the lhs. Also, /; can be placed on the rhs, immediately after the (compound) expression.

```
s[x_?EvenQ] := N[Sqrt[x]]


s[6]


2.44949


s[5]


s[5]
```

Localizing Names in a Rewrite Rule

As we have pointed out, when the rhs of a `Set` or `SetDelayed` function is evaluated (which occurs when a `Set` function is first entered and when a `SetDelayed` rewrite rule is first called), rewrite rules for all of its auxillary functions are placed in the global rule base. This can cause a problem if a name being used in a program conflicts with the use of the name elsewhere.

We can prevent a name clash by "insulating" the auxillary functions within the rewrite rule so that they are not placed in the global rule base as separate rewrite rules; they will only "exist" while being used in the evaluation of the rule.

This is usually done using the `Module` function.

```
lhs :=
  Module[{name1 = val1, name2, ...}, rhs]


    t[y_] := Module[{m}, m = 2; y + m]


?m


Global'm
```

```
t[3]
```

```
7
```

```
?m
```

```
Global'm
```

Creating Rewrite Rules Dynamically

One way to speed up the evaluation of an expression in *Mathematica* is to get it to remember the values it computes. This can be done by creating Set rewrite rules during the evaluation of a SetDelayed rewrite rule. To do this, a SetDelayed function is written with a rhs that is a Set function of the same name.

```
f[x_] := f[x] = rhs
```

An example of this is a program that computes Fibonacci numbers.

```
fib[0] := 0
fib[1] := 1
fib[n_] := fib[n] = fib[n - 1] + fib[n - 2]
```

When these rewrite rules are entered, the three rewrite rules are placed in the global rule base.

```
?fib
```

```
Global'fib
```

```
fib[0] := 0
```

```
fib[1] := 1
```

```
fib[n_] := fib[n] = fib[n - 1] + fib[n - 2]
```

When the program is subsequently run for a particular value of n, e.g., 3

```
fib[3]
```

```
2
```

the final value of fib[3] and the intermediate value of fib[2] are both computed. As these values are determined, they are entered as rewrite rules into the global rule base as we can see by querying the global rule base.

```
?fib
```

```
Global'fib
```

```
fib[0] := 0
```

```
fib[1] := 1
```

```
fib[2] = 1
```

```
fib[3] = 2
```

```
fib[n_] := fib[n] = fib[n - 1] + fib[n - 2]
```

The main benefit of this method is that having these additional rules in the global rule base saves computing time later because the new rewrite rules can be used when applicable, obviating the need to recalculate previously determined values. The use of dynamic programming is especially useful for recursive programs.

Ordering Rewrite Rules

When the lhs of more than one built-in and/or user-defined rewrite rule is found to pattern-match an expression (which occurs when the lhs's only differ in their specificity), the choice of which rule to use is determined by the order of precedence:

A user-defined rule is used before a built-in rule.

A more specific rule is used before a more general rule (a rule is more specific, the fewer expressions it pattern-matches).

So, for example, if we have two rewrite rules whose lhs's have the same name but whose labeled patterns have different specificity, both rules will appear in the global rule base (since their lhs 's are not identical) and the more specific rule will be used in preference to the more general rule. For example, if we enter both of the following function definitions

```
f[x_] := x^2
```

```
f[x_Integer] := x^3
```

and then query the rule base,

```
?f
```

Global'f

f[x_Integer] := x^3

f[x_] := x^2

Now, entering f with an real-valued argument

```
f[6.]
```

    36.

returns a different result from entering f with an integer-valued argument

```
f[6]
```

    216

because, while an integer-valued argument pattern-matches both x_ and x_Integer and hence pattern-matches both of the f rewrite rules, the second rule is a more specific pattern-match for the integer value 6.

Note: If *Mathematica* cannot decide which rule is more general, it uses the rules in the order in which they appear in the global rule base.

The ordering of rewrite rules makes it possible for us to create sets of rewrite rules with the same name that give different results, depending on the arguments used. This is key to writing rule-based programs.

Associating Rewrite Rules with Symbols

When a rewrite rule is placed in the global rule base, it is associated with (or attached to) the leftmost symbol on the lhs of the rewrite rule (the definition is said to be a downvalue of the associated symbol). For example,

```
s[t[y_]] := rhs[y]

?s
```

Global's

```
s[t[y_]] := rhs[y]

?t
```

Global't

To associate a rule with a symbol other than the leftmost symbol, the Upvalue function can be used (the definition is said to be an upvalue of the associated symbol). For example,

```
u/: v[u[z_]] := rhs[z]

?v
```

Global'v

```
?u
```

Global'u

```
v[u[y_]] ^:= rhs[y]
```

When there is a choice as to which symbol to associate with a rewrite rule, it is preferable to use the less-common symbol.

Note: It is necessary to be careful about the labeling of patterns in rewrite rules because if the lhs's of two or more rules are identical except for the labeling, these rules will all be placed in the global rule base and it is uncertain which rule will be used. For example,

```
w[y_] := y^3

w[x_] := x^4

w[2]
```

8

```
?w
```

```
Global'w
```

```
w[y_] := y^3
```

```
w[x_] := x^4
```

Transformation Rules

There are times when we want a rewrite rule to only be applied to (i.e., used inside) a specific expression, rather than being placed in the global rule base where it will be used whenever it pattern-matches an expression. For example, the 'temporary' substitution of a value for a name in an expression may be preferable to the permanent assignment of the name to the value via a Set function. When this is the case, the ReplaceAll function can be used together with a Rule or RuleDelayed function to create a transformation (or local rewrite) rule which is placed directly after the expression to which it is to be applied.

Using the Rule (->) Function

A Rule function is attached to an expression. It is written

$$\text{expression /. lhs -> rhs}$$

The lhs can be written using symbols, numbers, or labeled patterns.

When an expression with an attached Rule transformation rule is entered, the expression itself is evaluated first. Then, both the lhs and rhs of the Rule transformation rule are evaluated. Finally, the fully evaluated transformation rule is used in the evaluated expression. For example, we can create a list

```
Table[x, {4}]/. x -> Random[Integer]
```

```
{1, 1, 1, 1}
```

Using the Trace function,

```
Trace[Table[x, {4}]/. x ->Random[Integer]]
```

```
{{Table[x, {4}], {x, x, x, x}},
  {{Random[Integer], 1}, x -> 1,
    x -> 1}, {x, x, x, x} /. x -> 1,
   {1, 1, 1, 1}}
```

we see that Random[Integer] was evaluated before it was substituted for x in the list and hence all of the elements in the list are identical.

Note: A transformation rule is applied only once to each part of an expression (in contrast to a rewrite rule).

As another example of a transformation rule, we can convert a sum of elements into a list of the elements:

```
ReleaseHold[Hold[1 + 2 + 3] /. Plus -> List]
```

```
{1, 2, 3}
```

We can attach a list of rules to an expression using

```
expression /. {lhs1 -> rhs1, lhs2 -> rhs2,..}
```

For example,

```
{a, b, c}/.{c -> b, b -> a}
```

```
{a, a, b}
```

Note: Multiple transformation rules are used in parallel. The rules are applied in order so that a later rule in the list is used only if all the earlier rules do not match, and only one transformation rule at most, is applied to a given part of an expression, and no matching rules are used thereafter.

Using the RuleDelayed (:>) Function

A RuleDelayed function is attached to an expression. It is written

```
expression /. lhs :> rhs
```

or, for a list of rules

```
expression /. {lhs1 :> rhs1, lhs2 :> rhs2,..}
```

The lhs can be written using symbols, numbers, or labeled patterns.

When an expression with an attached rule is entered, the expression itself is evaluated first. Then, the lhs of the RuleDelayed transformation rule is evaluated but the rhs is not evaluated. Finally, the partially evaluated transformation rule is used in the evaluated expression (the unevaluated rhs will be evaluated subsequently).

For example,

```
Table[x, {4}]/. x :> Random[Integer]
```

```
{1, 1, 0, 1}
```

Using the Trace function,

```
Trace[Table[x, {4}]/. x :> Random[Integer]]
```

```
{{Table[x, {4}], {x, x, x, x}},
  {x, x, x, x} /. x :> Random[Integer],
  {Random[Integer], Random[Integer],
   Random[Integer], Random[Integer]},
  {Random[Integer], 0},
  {Random[Integer], 1},
  {Random[Integer], 0},
  {Random[Integer], 1}, {0, 1, 0, 1}}
```

we see that the unevaluated expression, Random[Integer], is first substituted for x in the list and then each occurrence of the expression in the list is evaluated, resulting in a list whose elements have varying values.

Placing Constraints on a Transformation Rule

By placing /; condition immediately after a RuleDelayed :> transformation rule, its use can be restricted in the same way that using /; condition can be used to restrict the use of a SetDelayed rewrite rule.

Applying a Transformation Rule Repeatedly

To apply one or more transformation rules repeatedly to an expression until the expression no longer changes, the ReplaceRepeated function is used. For example,

```
{a, b, c}//.{c -> b, b -> a}
```

```
{a, a, a}
```

Note: In using //. with a list of transformation rules, it is important to keep in mind the order of application of the rules. The transformation rules are not repeatedly applied in order. Rather, each rule, in turn, is applied repeatedly.

Functional Programming Style

Working with Functions

Mathematica works with built-in and user-defined functions in ways that are characteristic of the "functional" style of programming.

Nested Function Calls

Consider the following consecutive computations:

```
Tan[4.0]
```

```
1.15782
```

```
Sin[%]
```

```
0.915931
```

```
Cos[%]
```

```
0.609053
```

We can combine these function calls into a nested function call.

```
Cos[Sin[Tan[4.0]]]
```

```
0.609053
```

Notice that the result of one function call is immediately fed into another function without having to first name (or declare) the result.

A nested function call is the application of a function to the result of applying another function to some argument value. In applying functions successively, it is not necessary to declare the value of the result of one function call prior to using it as an argument in another function call.

We can illustrate the use of nested function calls using a deck of playing cards:

```
(* creating a deck of cards *)
```

```
Range[2, 10]
```

```
{2, 3, 4, 5, 6, 7, 8, 9, 10}
```

```
Join[%, {J, Q, K, A}]

{2, 3, 4, 5, 6, 7, 8, 9, 10, J, Q, K, A}

Outer[List, {c, d, h, s}, %]

{
 {{c, 2}, {c, 3}, {c, 4}, {c, 5}, {c, 6},
  {c, 7}, {c, 8}, {c, 9}, {c, 10}, {c, J},
  {c, Q}, {c, K}, {c, A}},

 {{d, 2}, {d, 3}, {d, 4}, {d, 5}, {d, 6},
  {d, 7}, {d, 8}, {d, 9}, {d, 10}, {d, J},
  {d, Q}, {d, K}, {d, A}},

 {{h, 2}, {h, 3}, {h, 4}, {h, 5}, {h, 6},
  {h, 7}, {h, 8}, {h, 9}, {h, 10}, {h, J},
  {h, Q}, {h, K}, {h, A}},

 {{s, 2}, {s, 3}, {s, 4}, {s, 5}, {s, 6},
  {s, 7}, {s, 8}, {s, 9}, {s, 10}, {s, J},
  {s, Q}, {s, K}, {s, A}} }

Flatten[%, 1]

{{c, 2}, {c, 3}, {c, 4}, {c, 5}, {c, 6},
 {c, 7}, {c, 8}, {c, 9}, {c, 10}, {c, J},
 {c, Q}, {c, K}, {c, A}, {d, 2}, {d, 3}, {d, 4},
 {d, 5}, {d, 6}, {d, 7}, {d, 8}, {d, 9}, {d, 10},
 {d, J}, {d, Q}, {d, K}, {d, A}, {h, 2}, {h, 3},
 {h, 4}, {h, 5}, {h, 6}, {h, 7}, {h, 8}, {h, 9},
 {h, 10}, {h, J}, {h, Q}, {h, K}, {h, A}, {s, 2},
 {s, 3}, {s, 4}, {s, 5}, {s, 6}, {s, 7}, {s, 8},
 {s, 9}, {s, 10}, {s, J}, {s, Q}, {s, K}, {s, A}}

cardDeck =
 Flatten[Outer[List,{c, d, h, s},Join[Range[2, 10], {J, Q, K, A}]], 1]

{{c, 2}, {c, 3}, {c, 4}, {c, 5}, {c, 6},
 {c, 7}, {c, 8}, {c, 9}, {c, 10}, {c, J},
 {c, Q}, {c, K}, {c, A}, {d, 2}, {d, 3}, {d, 4},
 {d, 5}, {d, 6}, {d, 7}, {d, 8}, {d, 9}, {d, 10},
 {d, J}, {d, Q}, {d, K}, {d, A}, {h, 2}, {h, 3},
 {h, 4}, {h, 5}, {h, 6}, {h, 7}, {h, 8}, {h, 9},
 {h, 10}, {h, J}, {h, Q}, {h, K}, {h, A}, {s, 2},
 {s, 3}, {s, 4}, {s, 5}, {s, 6}, {s, 7}, {s, 8},
 {s, 9}, {s, 10}, {s, J}, {s, Q}, {s, K}, {s, A}}
```

(* shuffling a deck of cards *)

```
Transpose[Sort[Transpose[{Table[ Random[], {52}], cardDeck}]]][[2]]
```

```
{{s, Q}, {s, 7}, {d, 5}, {c, 6}, {c, K}, {c, 3},
 {c, 4}, {h, J}, {d, Q}, {c, 2}, {c, 8}, {s, A},
 {c, Q}, {d, 2}, {h, Q}, {h, 5}, {c, 5}, {c, 7},
 {s, 4}, {h, 4}, {c, 10}, {d, 4}, {s, 8}, {h, 7},
 {d, 10}, {c, 9}, {s, 6}, {c, A},{s, 10}, {d, 3},
 {d, 8}, {d, 7}, {h, K}, {s, K}, {d, A}, {h, A},
 {s, 3}, {h, 3}, {h, 2}, {s, J}, {d, 9}, {d, 6},
 {d, K}, {h, 10}, {d, J}, {h, 8}, {s, 2}, {s, 9},
 {h, 9}, {h, 6}, {s, 5}, {c, J}}
```

Another example changes the letters of a word from lowercase to uppercase.

```
FromCharacterCode[ToCharacterCode["darwin"] - 32]
```

```
DARWIN
```

Anonymous Functions

User-defined anonymous functions can be created and used "on the spot" without being named or entered prior to being used.

An anonymous function is written using the same form as the rhs of a rewrite rule, replacing variable symbols with #1, #2,... and enclosing the expression in parentheses followed by an ampersand.

This notation can be demonstrated by converting some simple user-defined functions into anonymous functions. For example, a rewrite rule that squares a value

```
square[x_] := x^2
```

can be written as an anonymous function and applied to an argument, e.g., 5, instantly

```
(#^2)&[5]
```

An example of an anonymous function with two arguments, raises the first argument to the power of the second argument.

```
(#1^#2)&[5, 3]
```

```
125
```

It is important to distinguish between an anonymous function that takes multiple arguments and an anonymous function that takes a list with multiple elements as its argument.

For example, the anonymous function just given doesn't work with an ordered pair argument

```
(#1^#2)&[{2, 3}]
```

```
Function::slotn:
                        #2
  Slot number 2 in #1    &
    cannot be filled from
      #2
    (#1    & )[{2, 3}].
```

If we want to perform the operation on the components of an ordered pair, the appropriate anonymous function is

```
(#[[1]]^#[[2]])&[{2, 3}]
```

```
8
```

Nesting Anonymous Functions

Anonymous functions can be nested, in which case it is sometimes necessary to use the form

```
Function[x, body]
```

```
Function[{x, y, ...}, body]
```

rather than the (.#..)& form, in order to distinguish between the arguments of the different anonymous functions.

```
(#^3)&[(# + 2)&[3]]
```

```
125
```

```
Function[y, y^3][Function[x, x + 2][3]]
```

```
125
```

The two forms can also be used together.

```
Function[y, y^3][(# + 2)&[3]]
```

```
125
```

```
(#^3)&[Function[x, x + 2][3]]
```

```
125
```

Anonymous functions are useful for making predicates and arguments for higher-order functions.

Note: An anonymous predicate function is written using the (.#..&) form.

Higher-Order Functions

A higher-order function takes a function as an argument and/or returns a function as a result. This is known as 'treating functions as first-class objects.' We'll illustrate the use of some of the most important built-in higher-order functions.

Apply

```
?Apply
```

```
Apply[f, expr] or f @@ expr replaces the head of expr by f.
Apply[f, expr, levelspec] replaces heads in parts of expr specified by
levelspec.
```

We have already seen Apply used to add the elements of a linear list. Given a nested list argument, Apply can be used on the outer list or the interior lists. For example, for a general function f and a nested list:

```
Apply[f, {{a, b}, {c, d}}]
```

```
f[{a, b}, {c, d}]
```

```
Apply[f, {{a, b}, {c, d}}, 2]
```

```
{f[a, b], f[c, d]}
```

Map

```
?Map
```

```
Map[f, expr] or f /@ expr applies f to each element on the first level
in expr.
Map[f, expr, levelspec] applies f to parts of expr specified by
levelspec.
```

For a general function f and a linear list

```
Map[f, {a, b, c, d}]
```

```
{f[a], f[b], f[c], f[d]}
```

For a nested list structure, Map can be applied to either the outer list or to the interior lists, or to both. For example, for a general function g,

```
Map[g, {{a, b},  {c, d}}]
```

```
{g[{a, b}], g[{c, d}]}
```

```
Map[g, {{a, b},  {c, d}}, {2}]
```

```
{{g[a], g[b]}, {g[c], g[d]}}
```

```
Map[g, {{a, b},  {c, d}}, 2]
```

```
{g[{g[a], g[b]}], g[{g[c], g[d]}]}
```

MapThread

```
?MapThread
```

```
MapThread[f, {{a1, a2, ...}, {b1, b2, ...}, ...}] gives
{f[a1, b1, ...], f[a2, b2, ...], ...}.
```

```
MapThread[f, {xa, xb, ...}, n] maps f over the nth level of the
n-dimensional tensors xa, xb, ... .
```

For a general function g and a nested list,

```
MapThread[g, {{a, b, c}, {x, y, z}}]
```

```
{g[a, x], g[b, y], g[c, z]}
```

```
MapThread[List, {{a, b ,c }, {x, y, z}}]
```

```
{{a, x}, {b, y}, {c, z}}
```

```
MapThread[Plus, {{a, b ,c }, {x, y, z}}]
```

```
{a + x, b + y, c + z}
```

```
MapThread[g, {{{a, b},{c, d}}, {{e, f}, {g, h}}}]
```

```
{g[{a, b}, {e, f}], g[{c, d}, {g, h}]}
```

```
MapThread[g, {{{a, b},{c, d}}, {{e, f}, {g, h}}}, 2]//TableForm
```

```
g[a, e]   g[b, f]
g[c, g]   g[d, h]
```

NestList and Nest

Nest performs a nested function call, applying the same function repeatedly.

The Nest operation applies a function to a value, then applies the function to the result, and then applies the function to that result and then applies . . . and so on a specified number of times.

```
?NestList
```

```
NestList[f, expr, n] gives a list of the results of applying f to expr
0 through n times.
```

```
{0.7, Sin[0.7], Sin[Sin[0.7]], Sin[Sin[Sin[0.7]]]}
```

```
{0.7, 0.644218, 0.600573, 0.565115}
```

```
NestList[Sin, 0.7, 3]
```

```
{0.7, 0.644218, 0.600573, 0.565115}
```

If we are only interested in the final result of the NestList operation, we can use the Nest function, which does not return the intermediate results.

```
?Nest
```

```
Nest[f, expr, n] gives an expression with f applied n times to expr.
```

FixedPointList and FixedPoint

The Nest operation does not stop until it has completed a specified number of function applications. There is another function that performs the Nest operation, stopping after whichever of the following occurs first: (a) there have been a specified number of function applications, (b) the result stops changing, or (c) some predicate condition is met.

```
?FixedPointList
```

FixedPointList[f, expr] generates a list giving the results of applying f repeatedly, starting with expr, until the results no longer change.

FixedPointList[f, expr, n] stops after at most n steps.

```
?FixedPoint
```

FixedPoint[f, expr] starts with expr, then applies f repeatedly until the result no longer changes.

FixedPoint[f, expr, n] stops after at most n steps.

As an example,

```
FixedPointList[Sin, 0.7, 5, SameTest -> (#2 < 0.65 &)]
```

{0.7, 0.644218}

```
FixedPointList[Sin, 0.7, 5, SameTest -> ((#1 - #2) < 0.045 &)]
```

{0.7, 0.644218, 0.600573}

Note: in these examples, #1 refers to the next-to-last element in the list being generated and #2 refers to the last element in the list.

FoldList and Fold

```
?FoldList
```

FoldList[f, x, {a, b, ...}] gives {x, f[x, a], f[f[x, a], b], ...}.

```
?Fold
```

Fold[f, x, list] gives the last element of FoldList[f, x, list].

The `Fold` operation takes a function, a value, and a list, applies the function to the value, and then applies the function to the result and the first element of the list, and then applies the function to the result and the second element of the list and so on. For example,

```
Fold[Plus, 0, {a, b, c ,d}]
```

```
a + b + c + d
```

```
FoldList[Plus, 0, {a, b, c ,d}]
```

```
{0, a, a + b, a + b + c, a + b + c + d}
```

Listability

Built-in functions are `Listable`, so that they are automatically applied in parallel to the elements of their list argument(s).

```
Log[{a, b, c, d}]
```

```
{Log[a], Log[b], Log[c], Log[d]}
```

```
Log[{{a, b}, {c, d}}]
```

```
{{Log[a], Log[b]}, {Log[c], Log[d]}}
```

```
Plus[{a, b}, {c, d}]
```

```
{a + c, b + d}
```

We can make a user-defined function `Listable`. For example, the general function *h* can be made `Listable`

```
h[{a, b, c, d}]
```

```
h[{a, b, c, d}]
```

```
Map[h, {a, b, c, d}]
```

```
{h[a], h[b], h[c], h[d]}
```

```
Attributes[h] = Listable;

h[{a, b, c, d}]

{h[a], h[b], h[c], h[d]}
```

Note that when a `Listable` function is applied to a nested list, it is applied "all the way down."

B | Working with Lists

Introduction

The greater the use you make of *Mathematica*'s built-in functions in your programs, the more efficient your programs will be (since the built-in functions are implemented in C). It is therefore to your advantage to have a large *Mathematica* vocabulary. For purposes of CA programming, the built-in functions for list manipulation are the most relevant. Here we will illustrate some of the uses of *Mathematica* functions for working with lists. The examples below are shown without explanation because the operation of these functions should become self-evident when you compare each input with its output (and you can, of course, make up your own lists and try them out with the built-in functions to see what happens).

Note: You can get more information about these built-in functions, including definitions and examples, by referring to the *Mathematica* book by Wolfram or, if you running *Mathematica*, by either entering ?name or ??name or by using the on-line function browser.

Creating a List

```
List[2.4, dog, Sin, {5, 3}, Pi,
    "Computer Simulations with Mathematica"]


{2.4, dog, Sin, {5, 3}, Pi, "Computer Simulations with Mathematica"}


Range[-4, 7, 3]


{-4, -1, 2, 5}


Range[4, 8]


{4, 5, 6, 7, 8}
```

```
Range[4]
```

```
{1, 2, 3, 4}
```

```
Range[1.5, 6.3, 0.75]
```

```
{1.5, 2.25, 3., 3.75, 4.5, 5.25, 6.}
```

```
Table[3 k, {k, 1, 10, 2}]
```

```
{3, 9, 15, 21, 27}
```

```
Table[i, {i, 1.5, 6.3, .75}]
```

```
{1.5, 2.25, 3., 3.75, 4.5, 5.25, 6.}
```

```
Table[3 i, {i, 2, 5}]
```

```
{6, 9, 12, 15}
```

```
Table[i^2, {i, 4}]
```

```
{1, 4, 9, 16}
```

```
Table[darwin, {4}]
```

```
{darwin, darwin, darwin, darwin}
```

```
Table[Random[], {4}]
```

```
{0.666124, 0.518974, 0.174427, 0.65537}
```

```
Table[{Random[], Random[]},{3}]
```

```
{{0.284749, 0.842514}, {0.307664, 0.416197},
  {0.646685, 0.216293}}
```

```
Table[i + j, {j, 1, 4}, {i, 1, 3}]
```

```
{{2, 3, 4}, {3, 4, 5}, {4, 5, 6}, {5, 6, 7}}
```

```
Table[i + j, {i, 1, 3},{j, 1, 4}]
```

```
{{2, 3, 4, 5}, {3, 4, 5, 6}, {4, 5, 6, 7}}
```

```
Table[i + j, {i, 1, 4}, {j, 1, 3}]//TableForm
```

```
2   3   4
3   4   5
4   5   6
5   6   7
```

```
Table[{Random[Integer], Random[Integer]}, {3}, {3}]
```

```
{{{1, 1}, {0, 0}, {0, 0}},
 {{0, 0}, {0, 0}, {1, 1}},
 {{1, 1}, {0, 0}, {1, 1}}}
```

```
Table[i + j, {i, 1, 3}, {j, 1, i}]
```

```
{{2}, {3, 4}, {4, 5, 6}}
```

```
Table[i + j, {i, 1, 3}, {j, 1, i}]//TableForm
```

```
2
3   4
4   5   6
```

```
Table[f[j], {j, 5}]
```

```
{f[1], f[2], f[3], f[4], f[5]}
```

```
Array[f, {5}]
```

```
{f[1], f[2], f[3], f[4], f[5]}
```

```
Table[f[1, j], {i, 4}, {j, 4}]
```

```
{{f[1, 1], f[1, 2], f[1, 3], f[1, 4]},
 {f[1, 1], f[1, 2], f[1, 3], f[1, 4]},
 {f[1, 1], f[1, 2], f[1, 3], f[1, 4]},
 {f[1, 1], f[1, 2], f[1, 3], f[1, 4]}}
```

```
Array[f, {3, 4}]
```

```
{{f[1, 1], f[1, 2], f[1, 3], f[1, 4]},
 {f[2, 1], f[2, 2], f[2, 3], f[2, 4]},
 {f[3, 1], f[3, 2], f[3, 3], f[3, 4]}}
```

Measuring a List

```
Length[{a, b, c, d, e, f}]
```

6

```
Length[{{{1, 2}, {3, 4}, {5, 6}},  {{a, b}, {c, d}, {e, f}}}]
```

2

```
Dimensions[{{{1, 2}, {3, 4}, {5, 6}},  {{a, b}, {c, d}, {e, f}}}]
```

{2, 3, 2}

Locating Elements in a List

```
Position[{5, 7, 5, 2, 1, 4}, 5]
```

{{1}, {3}}

```
Position[{{a, b, c}, {d, e, f}}, f]
```

{{2, 3}}

Extracting Elements from a List

```
Part[{2, 3, 7, 8, 1, 4}, 3]
```

7

```
{2, 3, 7, 8, 1, 4}[[3]]
```

7

```
{{1, 2}, {a, b}, {3, 4}, {c, d}, {5, 6},{e, f}}[[2, 1]]
```

a

```
{{1, 2}, {a, b}, {3, 4}, {c, d}, {5, 6},{e, f}}[[{2, 1}]]
```

{{a, b}, {1, 2}}

```
Take[{2, 3, 7, 8, 1, 4}, 2]

{2, 3}

Take[{2, 3, 7, 8, 1, 4}, -2]

{1, 4}

Take[{2, 3, 7, 8, 1, 4}, {2, 4}]

{3, 7, 8}

Take[{2, 3, 7, 8, 1, 4}, { -5, -3}]

{3, 7, 8}

Take[{2, 3, 7, 8, 1, 4}, { -4, 5}]

{7, 8, 1}

Drop[{2, 3, 7, 8, 1, 4}, 2]

{7, 8, 1, 4}

Drop[{2, 3, 7, 8, 1, 4}, -2]

{2, 3, 7, 8}

Drop[{2, 3, 7, 8, 1, 4}, {3, 5}]

{2, 3, 4}

Delete[{2, 3, 7, 8, 1, 4}, 2]

{2, 7, 8, 1, 4}

Delete[{2, 3, 7, 8, 1, 4}, {{2}, {5}}]

{2, 7, 8, 4}

Select[{2, 3, 7, 8, 1, 4}, EvenQ]

{2, 8, 4}
```

```
Cases[{2, {3, 7}, 8, d}, {__}]

{{3, 7}}

DeleteCases[{2, {3, 7}, 8}, x_List]

{2, 8}

First[{2, 3, 7, 8, 1, 4}]

2

Last[{2, 3, 7, 8, 1, 4}]

4

Rest[{2, 3, 7, 8, 1, 4}]

{3, 7, 8, 1, 4}
```

Rearranging a List

```
Reverse[{2, 7, e, 1, a, 5}]

{5, a, 1, e, 7, 2}

Sort[{2, 7, e, 1, a, 5}]

{1, 2, 5, 7, a, e}

Sort[{{2, 6}, {7, b, d}, {7, f, a}, {2, 3.5}, {7, v}}]

{{2, 3.5}, {2, 6}, {7, v}, {7, b, d},{7, f, a}}

RotateLeft[{2, 7, e, 1, a, 5}, 2]

{e, 1, a, 5, 2, 7}

RotateRight[{2, 7, e, 1, a, 5}, 2]

{a, 5, 2, 7, e, 1}

RotateRight[{2, 7, e, 1, a, 5}, -2]

{e, 1, a, 5, 2, 7}
```

Nesting and Unnesting Lists

```
Flatten[{{{3, 1}, {2, 4}}, {{5,3}, {7, 4}}}]
```

```
{3, 1, 2, 4, 5, 3, 7, 4}
```

```
Flatten[{{{3, 1}, {2, 4}}, {{5,3}, {7, 4}}}, 1]
```

```
{{3, 1}, {2, 4}, {5, 3}, {7, 4}}
```

```
Partition[{2, 3, 7, 8, 1, 4}, 2]
```

```
{{2, 3}, {7, 8}, {1, 4}}
```

```
Partition[{2, 3, 7, 8, 1, 4}, 1, 2]
```

```
{{2}, {7}, {1}}
```

```
Partition[{2, 3, 7, 8, 1, 4}, 2, 1]
```

```
{{2, 3}, {3, 7}, {7, 8}, {8, 1}, {1, 4}}
```

```
Transpose[{{x1, x2 ,x3, x4}, {y1, y2, y3, 44}}]
```

```
{{x1, y1}, {x2, y2}, {x3, y3},
 {x4, 44}}
```

```
Transpose[{{x1, y1}, {x2, y2}, {x3, y3}, {x4, y4}}]
```

```
{{x1, x2, x3, x4}, {y1, y2, y3, y4}}
```

```
Transpose[{{5, 2 ,7, 3}, {4, 6, 8, 4}, {6, 5, 3, 1}}]
```

```
{5, 2, 3, 7, 8, 1, 4}
```

Adding Elements to a List

```
Prepend[{2, 3, 7, 8, 1, 4}, 5]
```

```
{5, 2, 3, 7, 8, 1, 4}
```

```
Append[{2, 3, 7 , 8, 1, 4}, 5]
```

```
{2, 3, 7, 8, 1, 4, 5}
```

```
Append[{2, 3, 7 , 8, 1, 4}, {7, w}]

{2, 3, 7, 8, 1, 4, {7, w}}

Join[{2, 5, 7, 3}, {d, a, e, j}]

{2, 5, 7, 3, d, a, e, j}

Union[{4, 1, 2}, {5, 1, 2}]

{1, 2, 4, 5}

Union[{4, 1, 2, 5, 1, 2}]

{1, 2, 4, 5}

Insert[{2, 3, 7, 8, 1, 4}, 5, 3]

{2, 3, 5, 7, 8, 1, 4}
```

Changing Elements in a List

```
ReplacePart[{2, 3, 7, 8, 1, 4}, 5, 2]

{2, 5, 7, 8, 1, 4}

ReplacePart[{2, 3, 7, 8, 1, 4}, 5, {{2}, {4}}]

{2, 5, 7, 5, 1, 4}
```

Treating Lists as Sets

```
Complement[{2, 8, 7, 4, 8}, {3, 5, 4}]

{2, 7, 8}

Intersection[{2, 8, 7, 4, 3}, {3, 5, 4}]

{3, 4}
```

C | Program Listing

The Game of Life

```
OblaDeOblaDa[n_, t_]:=
Module[{initConfig, Moore, update, LiveConfigs, DieConfigs},

 initConfig = Table[Random[Integer],{n},{n}] ;

 LiveConfigs =
   Join[Map[Join[{0}, #]&, Permutations[{1, 1, 1, 0, 0, 0, 0, 0}]],
        Map[Join[{1}, #]&, Permutations[{1, 1, 1, 0, 0, 0, 0, 0}]],
        Map[Join[{1}, #]&, Permutations[{1, 1, 0, 0, 0, 0, 0, 0}]]];

 DieConfigs =
   Complement[Flatten[
     Map[Permutations, Map[Join[Table[1, {#}], Table[0, {(9 - #)}]]&,
         Range[0, 9]]], 1], LiveConfigs];

 Apply[(update[##] = 1)&, LiveConfigs, 1];
 Apply[(update[##] = 0)&, DieConfigs, 1];

 Moore[func__, lat_] :=
  MapThread[func, Map[RotateRight[lat, #]&,
           {{0, 0}, {1, 0}, {0, -1}, {-1, 0}, {0, 1},
            {1, -1}, {-1, -1}, {-1, 1}, {1, 1}}], 2];

 FixedPointList[Moore[update, #]&, initConfig, t]
]

LifeGoesOn[n_, t_]:=
Module[{initConfig, Moore, update},

 initConfig = Table[Random[Integer],{n},{n}] ;

 update[1, 3] := 1;
 update[0, 3] := 1;
 update[1, 4] := 1;
 update[_, _] := 0;
```

```
Moore[func__, lat_] :=
 MapThread[func, Map[RotateRight[lat, #]&,
           {{0, 0}, {1, 0}, {0, -1}, {-1, 0}, {0, 1},
            {1, -1}, {-1, -1}, {-1, 1}, {1, 1}}], 2]

FixedPointList[MapThread[update, {#, Moore[Plus, #]}, 2]&,
               initConfig, t]
]
```

Maze-Solving

```
PathToEnlightenment[ maze_] :=
  Module[ {mazeSolve, VonNeumann},

    mazeSolve[0, 1, 1, 1, 0]   := 1;
    mazeSolve[0, 1, 1, 0, 1]   := 1;
    mazeSolve[0, 1, 0, 1, 1]   := 1;
    mazeSolve[0, 0, 1, 1, 1]   := 1;
    mazeSolve[0, 1, 1, 1, 1]   := 1;
    mazeSolve[x_, _, _, _, _] := x;

    VonNeumann[func__, lat_] :=
       MapThread[func, Map[RotateRight[lat, #]&,
                 {{0, 0}, {1, 0}, {0, -1}, {-1, 0}, {0, 1}}]], 2];

    FixedPoint[VonNeumann[mazeSolve, #]&, maze]
  ]
```

Traffic Flow

```
keepOnTrucking[n_, p_, t_] :=
 Module[{road, drive},

   road = Table[Floor[p + Random[]],{2},{n}];

   drive[1, 0, _, _, _, _] =  0;
   drive[1, 1, _, 0, _, 0] =  0;
   drive[1, 1, _, 1, _, _] =  1;
   drive[0, _, 1, _, _, _] =  1;
   drive[0, _, 0, 1, 1, _] =  1;
   drive[x_, _, _, _, _, _] := x;

   NestList[MapThread[drive,
               {#, RotateRight[#, {0, -1}], RotateRight[#, {0, 1}],
                RotateRight[#, {1, 0}],  RotateRight[#, {1, -1}],
                RotateRight[#, {1, 1}]}, 2]&, road, t]
 ]
```

```
indy500[n_, p_, t_] :=
  Module[{roadWithObstacle, driveWithObstacle},

    roadWithObstacle =
      ReplacePart[Table[Floor[p + Random[]],{2}, {n}],  c,
                  {Random[Integer,{1, 2}], Random[Integer, {1, n}]}];

    driveWithObstacle[1, 0, _, _, _, _] =  0;
    driveWithObstacle[1, 1 | c, _, 0, _, 0 | c] =  0;
    driveWithObstacle[1, 1 | c, _, 1 | c, _, _] =  1;
    driveWithObstacle[0, _, 1, _, _, _] =  1;
    driveWithObstacle[0, _, 0 | c, 1, 1 | c, _] =  1;
    driveWithObstacle[x_, _, _, _, _, _] := x;

    NestList[MapThread[driveWithObstacle,
                  {#, RotateRight[#, {0, -1}], RotateRight[#, {0, 1}],
                   RotateRight[#, {1, 0}],  RotateRight[#, {1, -1}],
                   RotateRight[#, {1, 1}]}, 2]&, roadWithObstacle, t]
  ]
```

Phase Ordering and Spinoidal Decomposition

```
PhaseOrderingNonConserved[d_, m_, t_]:=
 Module[{initconfig, Moore, separate, newlat},

  initconfig = Table[Random[Real, {-0.1, 0.1}], {m}, {m}];

  separate[x_, n_, e_, s_, w_, ne_, se_, sw_, nw_] :=
    1.3 Tanh[x] + d ((n + e + s + w)/6 + (ne + se + sw + nw)/12 - x);

  Moore[func__, lat_] :=
    MapThread[func, Map[RotateRight[lat, #]&,
           {{0, 0}, {1, 0}, {0, -1}, {-1, 0}, {0, 1},
            {1, -1}, {-1, -1}, {-1, 1}, {1, 1}}], 2];

  newlat = Nest[Moore[separate, #]&, initconfig, t];

  ListDensityPlot[newlat, Mesh -> False, FrameTicks -> None]
 ]
```

```
PhaseOrderingConserved[d_, m_,  t_] :=
 Module[{initconfig, nnave, Itn, newlat, Moore},

   initconfig = Table[Random[Real, {-0.1, 0.1}], {m}, {m}];

   nnave[x_, n_, e_, s_, w_, ne_, se_, sw_, nw_] :=
                        (n + e + s + w)/6 + (ne + se + sw + nw)/12;

   Moore[func__, lat_] :=
     MapThread[func, Map[RotateRight[lat, #]&,
               {{0, 0}, {1, 0}, {0, -1}, {-1, 0}, {0, 1},
                {1, -1}, {-1, -1}, {-1, 1}, {1, 1}}], 2];

   Itn[mat_] := d (Moore[nnave, mat] - mat) + 1.3 Tanh[mat] - mat;

   newlat = Nest[Function[y, (# + y - Moore[nnave, y])][Itn[#]]&,
                 initconfig, t];

   ListDensityPlot[newlat, Mesh -> False, FrameTicks -> None]
 ]


Vote[s_, t_]:=
 Module[{rule, init, Moore, newlat},

   init = Table[Random[Integer], {s}, {s}];

   Moore[func__, lat_] :=
     MapThread[func, Map[RotateRight[lat, #]&,
               {{0, 0}, {1, 0}, {0, -1}, {-1, 0}, {0, 1},
                {1, -1}, {-1, -1}, {-1, 1}, {1, 1}}], 2]

   rule[4] := 1;
   rule[5] := 0;
   rule[x_] := Floor[x/5];
   Attributes[rule] = Listable;

   newlat = Nest[rule[Moore[Plus, #]]&, init, t];

   ListDensityPlot[newlat, Mesh -> False, FrameTicks -> None]
 ]
```

Dendrite Formation

```
dendrite[D_Real, m_, lambda_, latheat_, delta_, undercool_, n_] :=
Module[{zerosDecorate, seedLat, phase, temp, Moore},

  zerosDecorate =
    Nest[(Prepend[Append[Map[Prepend[Append[#, {0, 0}], {0, 0}]&, #],
                    Table[{0, 0}, {Length[#] + 2}]],
              Table[{0, 0}, {Length[#] + 2}]])&, #, m]&;

  seedLat =
      zerosDecorate[Table[{1, undercool} Random[Integer], {5}, {5}]];

  phase[{0, a_}, {s_, _}, {t_, _}, {u_, _},
                {v_, _}, {w_, _}, {x_, _}, {y_, _}, {z_, _}] :=
      {1, a + latheat} /; MatchQ[1, s | t | u | v] &&
              a < undercool (1 + delta Random[Real, {-1, 1}]) +
                  lambda (2 (s + t + u + v) + (w + x + y + z) - 6);

  phase[{r_, a_}, {_, _}, {_, _}, {_, _}, {_, _},
                    {_, _}, {_, _}, {_, _}, {_, _}] := {r, a};

  temp[{r_, a_}, {_, b_}, {_, c_}, {_, d_},
        {_, e_}, {_, f_}, {_, g_}, {_, h_}, {_, i_}] :=
        {r, a + (D/m)*((b + c + d + e)/6+ (f + g + h + i)/12 - a)};

  Moore[func__, lat_] :=
    MapThread[func, Map[RotateRight[lat, #]&,
            {{0, 0}, {1, 0}, {0, -1}, {-1, 0}, {0, 1},
              {1, -1}, {-1, -1}, {-1, 1}, {1, 1}}], 2];

  FixedPoint[Nest[Moore[temp, #]&, zerosDecorate[Moore[phase, #]], m]&,
        seedLat, n]
]
```

Snowflake

```
snow[m_?EvenQ, n_?EvenQ, t_Integer] :=
Module[{seed, OddEven, snowflake, crystallize},

  seed = ReplacePart[Table[0, {m}, {n}], 1,
                {Ceiling[m/2], Ceiling[n/2]}];
```

```
snowflake[a_?OddQ, 0, 1, 0, 0, 0, 0, 0, _, _] := 1;
snowflake[a_?OddQ, 0, 0, 1, 0, 0, 0 ,0, _, _] := 1;
snowflake[a_?OddQ, 0, 0, 0, 1, 0, 0 ,0, _, _] := 1;
snowflake[a_?OddQ, 0, 0, 0, 0, 1, 0 ,0, _, _] := 1;
snowflake[a_?OddQ, 0, 0, 0, 0, 0, 1 ,0, _, _] := 1;
snowflake[a_?OddQ, 0, 0, 0, 0, 0, 0 ,1, _, _] := 1;

snowflake[a_?EvenQ, 0, 1, 0, 0, 0, _, _, 0, 0] := 1;
snowflake[a_?EvenQ, 0, 0, 1, 0, 0, _, _, 0, 0] := 1;
snowflake[a_?EvenQ, 0, 0, 0, 1, 0, _, _, 0, 0] := 1;
snowflake[a_?EvenQ, 0, 0, 0, 0, 1, _, _, 0, 0] := 1;
snowflake[a_?EvenQ, 0, 0, 0, 0, 0, _, _, 1, 0] := 1;
snowflake[a_?EvenQ, 0, 0, 0, 0, 0, _, _, 0, 1] := 1;

snowflake[_, b_, _, _, _, _, _, _, _, _] := b;

Attributes[snowflake] = Listable;

OddEven = Table[i,{i, m}, {n}];

crystallize =
   snowflake[OddEven, #,
             RotateRight[#, {-1, 0}], RotateRight[#, {1, 0}],
             RotateRight[#, {0, 1}],  RotateRight[#, {0, -1}],
             RotateRight[#, {1, 1}],  RotateRight[#, {-1, 1}],
             RotateRight[#, {1, -1}], RotateRight[#, {-1, -1}]]&;

NestList[crystallize, seed, t]
]
```

Single Walker

```
Meander[n_, t_] :=
 Module[{RND, walk, VonNeumann, initConf},

  RND := Random[Integer, {1, 4}];

  initConf = ReplacePart[Table[0, {2n + 1}, {2n + 1}], RND,
                      {n + 1, n + 1}];

  walk[1, 0, 0, 0, 0] := 0;
  walk[2, 0, 0, 0, 0] := 0;
  walk[3, 0, 0, 0, 0] := 0;
  walk[4, 0, 0, 0, 0] := 0;
```

```
      walk[0, 3, 0, 0, 0] := RND;
      walk[0, 0, 4, 0, 0] := RND;
      walk[0, 0, 0, 1, 0] := RND;
      walk[0, 0, 0, 0, 2] := RND;

      walk[0, 1, 0, 0, 0] := 0;
      walk[0, 2, 0, 0, 0] := 0;
      walk[0, 4, 0, 0, 0] := 0;
      walk[0, 0, 1, 0, 0] := 0;
      walk[0, 0, 2, 0, 0] := 0;
      walk[0, 0, 3, 0, 0] := 0;
      walk[0, 0, 0, 2, 0] := 0;
      walk[0, 0, 0, 3, 0] := 0;
      walk[0, 0, 0, 4, 0] := 0;
      walk[0, 0, 0, 0, 1] := 0;
      walk[0, 0, 0, 0, 3] := 0;
      walk[0, 0, 0, 0, 4] := 0;

      walk[0, 0, 0, 0, 0] := 0;

      VonNeumann[func__, lat_] :=
        MapThread[func, Map[RotateRight[lat, #]&,
                 {{0, 0}, {1, 0}, {0, -1}, {-1, 0}, {0, 1}}]], 2];

      NestList[VonNeumann[walk, #]&, initConf, t]
      ]
```

Shy Walkers

```
      Hermits[n_, p_, t_] :=
       Module[{walk, initConf, RND, MvonN},

         initConf = Table[Floor[p + Random[]], {n}, {n}] *
                    Table[Random[Integer, {1, 4}], {n}, {n}];

         RND := Random[Integer, {1, 4}];

         walk[1, 0, _, _, _, 4, _, _, _, _, _, _, _] := RND;
         walk[1, 0, _, _, _, _, _, _, 2, _, _, _, _] := RND;
         walk[1, 0, _, _, _, _, _, _, _, 3, _, _, _] := RND;
         walk[1, 0, _, _, _, _, _, _, _, _, _, _, _] := 0;
         walk[2, _, 0, _, _, 3, _, _, _, _, _, _, _] := RND;
         walk[2, _, 0, _, _, _, 1, _, _, _, _, _, _] := RND;
         walk[2, _, 0, _, _, _, _, _, _, 4, _, _, _] := RND;
         walk[2, _, 0, _, _, _, _, _, _, _, _, _, _] := 0;
```

```
walk[3, _, _, 0, _, _, 4, _, _, _, _, _, _] := RND;
walk[3, _, _, 0, _, _, _, 2, _, _, _, _, _] := RND;
walk[3, _, _, 0, _, _, _, _, _, _, _, 1, _] := RND;
walk[3, _, _, 0, _, _, _, _, _, _, _, _, _] := 0;
walk[4, _, _, _, 0, _, _, 1, _, _, _, _, _] := RND;
walk[4, _, _, _, 0, _, _, _, 3, _, _, _, _] := RND;
walk[4, _, _, _, 0, _, _, _, _, _, _, _, 2] := RND;
walk[4, _, _, _, 0, _, _, _, _, _, _, _, _] := 0;
walk[_?Positive, _, _, _, _,  _, _, _, _, _, _, _, _] := RND;
walk[0, 3, 4, _, _, _, _, _, _,  _, _, _, _] := 0;
walk[0, 3, _, 1, _, _, _, _, _,  _, _, _, _] := 0;
walk[0, 3, _, _, 2, _, _, _, _,  _, _, _, _] := 0;
walk[0, _, 4, 1, _, _, _, _, _,  _, _, _, _] := 0;
walk[0, _, 4, _, 2, _, _, _, _,  _, _, _, _] := 0;
walk[0, _, _, 1, 2, _, _, _, _,  _, _, _, _] := 0;
walk[0, 3, _, _, _, _, _, _, _,  _, _, _, _] := RND;
walk[0, _, 4, _, _, _, _, _, _,  _, _, _, _] := RND;
walk[0, _, _, 1, _, _, _, _, _,  _, _, _, _] := RND;
walk[0, _, _, _, 2, _, _, _, _,  _, _, _, _] := RND;
walk[0, _, _, _, _, _, _, _, _,  _, _, _, _] := 0;


MvonN[func__, lat_] :=
  MapThread[func, Map[RotateRight[lat, #]&,
           {{0, 0}, {1, 0}, {0, -1}, {-1, 0}, {0, 1},
            {1, -1}, {-1, -1}, {-1, 1}, {1, 1},
            {2, 0}, {0, -2}, {-2, 0}, {0, 2}}], 2];


NestList[MvonN[walk, #]& ,  initConf, t]
]
```

Courteous Walkers

```
Gentlemen[n_, p_, t_]:=
Module[{initConf, walk, rnd, MvonN},

rnd := {Random[Integer, {1, 4}], Random[]};

initConf = Table[rnd* Floor[Random[] + p], {n}, {n}];

walk[{1, a_}, {0, 0}, _, _, _, {4, b_},
     _, _, {2, c_}, {3, d_}, _, _, _] := rnd /; a != Max[a, b, c, d];
walk[{1, a_}, {0, 0}, _, _, _, {4, b_}, _, _, {2, c_}, _, _, _, _] :=
                                      rnd /; a != Max[a, b, c];
```

```
walk[{1, a_}, {0, 0}, _, _, _, {4, b_}, _, _, _, {3, d_}, _, _, _] :=
                                        rnd /; a != Max[a, b, d];
walk[{1, a_}, {0, 0}, _, _, _, _, _, _, {2, c_}, {3, d_}, _, _, _] :=
                                        rnd /; a != Max[a, c, d];
walk[{1, a_}, {0, 0}, _, _, _, {4, b_}, _,  _, _, _, _, _, _] :=
                                        rnd /; a != Max[a, b];
walk[{1, a_}, {0, 0}, _, _, _, _,  _, _, {2, c_}, _, _, _, _] :=
                                        rnd /; a != Max[a, c];
walk[{1, a_}, {0, 0}, _, _, _, _,  _, _, _, {3, d_}, _, _, _] :=
                                        rnd /; a != Max[a, d];
walk[{1, a_}, {0, 0}, _, _, _, _, _, _, _, _, _, _, _] := {0, 0};
walk[{2, a_}, _, {0, 0}, _, _, {3, b_},
     {1, c_}, _, _, _, {4, d_}, _, _] := rnd /; a != Max[a, b, c, d];
walk[{2, a_}, _, {0, 0}, _, _, {3, b_}, {1, c_}, _, _, _, _, _, _] :=
                                        rnd /; a != Max[a, b, c];
walk[{2, a_}, _, {0, 0}, _, _, {3, b_}, _, _, _, _, {4, d_}, _, _] :=
                                        rnd /; a != Max[a, b, d];
walk[{2, a_}, _, {0, 0}, _, _, _, {1, c_}, _, _, _, {4, d_}, _, _] :=
                                        rnd /; a != Max[a, c, d];
walk[{2, a_}, _, {0, 0}, _, _, {3, b_}, _, _, _, _, _, _, _] :=
                                        rnd /; a != Max[a, b];
walk[{2, a_}, _, {0, 0}, _, _, _, {1, c_}, _, _, _, _, _] :=
                                        rnd /; a != Max[a, c];
walk[{2, a_}, _, {0, 0}, _, _, _, _, _, _, _, {4, d_}, _, _] :=
                                        rnd /; a != Max[a, d];
walk[{2, a_}, _, {0, 0}, _, _, _, _, _, _, _, _, _, _] := {0, 0};
walk[{3, a_}, _, _, {0, 0}, _, _, {4, b_},
        {2, c_}, _, _, _, {1, d_}, _] := rnd /; a != Max[a, b, c, d];
walk[{3, a_}, _, _, {0, 0}, _, _, {4, b_}, {2, c_}, _, _, _, _, _] :=
                                        rnd /; a != Max[a, b, c];
walk[{3, a_}, _, _, {0, 0}, _, _, {4, b_}, _, _, _, _, {1, d_}, _] :=
                                        rnd /; a != Max[a, b, d];
walk[{3, a_}, _, _, {0, 0}, _, _,_, {2, c_}, _, _, _, {1, d_}, _] :=
                                        rnd /; a != Max[a, c, d];
walk[{3, a_}, _, _, {0, 0}, _, _, {4, b_}, _, _, _, _, _, _] :=
                                        rnd /; a != Max[a, b];
walk[{3, a_}, _, _, {0, 0}, _, _, _, {2, c_}, _, _, _, _, _] :=
                                        rnd /; a != Max[a, c];
walk[{3, a_}, _, _, {0, 0}, _, _, _, _, _, _, _, {1, d_}, _] :=
                                        rnd /; a != Max[a, d];
walk[{3, a_}, _, _, {0, 0}, _, _, _, _, _, _, _, _, _] := {0, 0};
walk[{4, a_}, _, _, _, {0, 0}, _, _, {1, b_},
        {3, c_}, _, _, _, {2, d_}] := rnd /; a != Max[a, b, c, d];
```

```
walk[{4, a_}, _, _, _, {0, 0}, _, _, {1, b_}, {3, c_}, _, _, _, _] :=
                                    rnd /; a != Max[a, b, c];
walk[{4, a_}, _, _, _, {0, 0}, _, _, {1, b_}, _, _, _, _, {2, d_}] :=
                                    rnd /; a != Max[a, b, d];
walk[{4, a_}, _, _, _, {0, 0}, _, _, _, {3, c_}, _, _, _, {2, d_}] :=
                                    rnd /; a != Max[a, c, d];
walk[{4, a_}, _, _, _, {0, 0}, _, _, {1, b_}, _, _, _, _, _] :=
                                        rnd /; a != Max[a, b];
walk[{4, a_}, _, _, _, {0, 0}, _, _, _, {3, c_}, _, _, _, _] :=
                                        rnd /; a != Max[a, c];
walk[{4, a_}, _, _, _, {0, 0}, _, _, _, _, _, _, _, {2, d_}] :=
                                        rnd /; a != Max[a, d];
walk[{4, a_}, _, _, _, {0, 0}, _, _, _, _, _, _, _, _] := {0, 0};
walk[{_?Positive, _}, _, _, _, _, _, _, _, _, _, _, _, _] := rnd;
walk[{0, 0}, {3, _}, _, _, _, _, _, _, _, _, _, _, _] := rnd;
walk[{0, 0}, _, {4, _}, _, _, _, _, _, _, _, _, _, _] := rnd;
walk[{0, 0}, _, _, {1, _}, _, _, _, _, _, _, _, _, _] := rnd;
walk[{0, 0}, _, _, _, {2, _}, _, _, _, _, _, _, _, _] := rnd;
walk[{0, 0}, _, _, _, _, _, _, _, _, _, _, _, _] := {0, 0};


MvonN[func__, lat_] :=
  MapThread[func, Map[RotateRight[lat, #]&,
            {{0, 0}, {1, 0}, {0, -1}, {-1, 0}, {0, 1},
             {1, -1}, {-1, -1}, {-1, 1}, {1, 1},
             {2, 0}, {0, -2}, {-2, 0}, {0, 2}}], 2];


NestList[MvonN[walk, #]&, initConf, t]
]
```

Interfacial Diffusion

```
interfacialDiffusion[n_, t_] :=
Module[{walk, initConf, RND, MvonN},

initConf = Join[Table[0, {n}, {n}],
                {Table[Random[Integer, {1, 4}], {n}]},
                {Table[b, {n}]}];

RND := Random[Integer, {1, 4}];

walk[1, 0, _, _, _, 4, _, _, _, _, _, _, _] := RND;
walk[1, 0, _, _, _, _, _, _, 2, _, _, _, _] := RND;
walk[1, 0, _, _, _, _, _, _, _, 3, _, _, _] := RND;
walk[1, 0, _, _, _, _, _, _, _, _, _, _, _] := 0;
walk[2, _, 0, _, _, 3, _, _, _, _, _, _, _] := RND;
```

```
        walk[2, _, 0, _, _, _, 1, _, _, _, _, _, _] := RND;
        walk[2, _, 0, _, _, _, _, _, _, _, 4, _, _] := RND;
        walk[2, _, 0, _, _, _, _, _, _, _, _, _, _] := 0;
        walk[3, _, _, 0, _, _, 4, _, _, _, _, _, _] := RND;
        walk[3, _, _, 0, _, _, _, 2, _, _, _, _, _] := RND;
        walk[3, _, _, 0, _, _, _, _, _, _, 1, _] := RND;
        walk[3, _, _, 0, _, _, _, _, _, _, _, _] := 0;
        walk[4, _, _, _, 0, _, _, 1, _, _, _, _, _] := RND;
        walk[4, _, _, _, 0, _, _, _, 3, _, _, _, _] := RND;
        walk[4, _, _, _, 0, _, _, _, _, _, _, _, 2] := RND;
        walk[4, _, _, _, 0, _, _, _, _, _, _, _, _] := 0;
        walk[_?Positive, _, _, _, _, _, _, _, _, _, _, _, _] := RND;
        walk[0, 3, 4, _, _, _, _, _, _, _, _, _, _] := 0;
        walk[0, 3, _, 1, _, _, _, _, _, _, _, _, _] := 0;
        walk[0, 3, _, _, 2, _, _, _, _, _, _, _, _] := 0;
        walk[0, _, 4, 1, _, _, _, _, _, _, _, _, _] := 0;
        walk[0, _, 4, _, 2, _, _, _, _, _, _, _, _] := 0;
        walk[0, _, _, 1, 2, _, _, _, _, _, _, _, _] := 0;
        walk[0, 3, _, _, _, _, _, _, _, _, _, _, _] := RND;
        walk[0, _, 4, _, _, _, _, _, _, _, _, _, _] := RND;
        walk[0, _, _, 1, _, _, _, _, _, _, _, _, _] := RND;
        walk[0, _, _, _, 2, _, _, _, _, _, _, _, _] := RND;
        walk[0, _, _, _, _, _, _, _, _, _, _, _, _] := 0;
        walk[b, _, _, _, _, _, _, _, _, _, _, _, _] := b;
        walk[1, 0, _, b, _, _, _, _, _, _, _, _, _] := RND;
        walk[1, b, _, _, _, _, _, _, _, _, _, _, _] := 0;


        MvonN[func__, lat_] :=
          MapThread[func, Map[RotateRight[lat, #]&,
                    {{0, 0}, {1, 0}, {0, -1}, {-1, 0}, {0, 1},
                     {1, -1}, {-1, -1}, {-1, 1}, {1, 1},
                     {2, 0}, {0, -2}, {-2, 0}, {0, 2}}], 2];


        NestList[MvonN[walk, #]&, initConf, t]
      ]
```

Driven Diffusion of Two Species

```
        DrivenDiffusion2Species[n_, p_, a_, t_]:=
        Module[{initPhase, driven, rndA, rndB},

        initPhase =
          Table[{{1, 2, 3, 4}[[Random[Integer, {1, 4}]]],
                 {A, B}[[Random[Integer, {1, 2}]]]} * Floor[Random[] + p],
                {n}, {n}];
```

```
rndA := {Random[Integer, {1, 2}],
         Random[Integer, {3, 4}]}[[2 -Floor[Random[] + a]]];

rndB := {Random[Integer, {1, 2}],
         Random[Integer, {3, 4}]}[[1 + Floor[Random[] + a]]];

driven[{1, A}, {x_?Positive, _}, _, _, _, _, _, _, _, _, _, _, _] :=
                                                            {rndA, A};
driven[{2, A}, _, {x_?Positive, _}, _, _, _, _, _, _, _, _, _, _] :=
                                                            {rndA, A};
driven[{3, A}, _, _, {x_?Positive, _}, _, _, _, _, _, _, _, _, _] :=
                                                            {rndA, A};
driven[{4, A}, _, _, _, {x_?Positive, _}, _, _, _, _, _, _, _, _] :=
                                                            {rndA, A};
driven[{1, B}, {x_?Positive, _}, _, _, _, _, _, _, _, _, _, _, _] :=
                                                            {rndB, B};
driven[{2, B}, _, {x_?Positive, _}, _, _, _, _, _, _, _, _, _, _] :=
                                                            {rndB, B};
driven[{3, B}, _, _, {x_?Positive, _}, _, _, _, _, _, _, _, _, _] :=
                                                            {rndB, B};
driven[{4, B}, _, _, _, {x_?Positive, _}, _, _, _, _, _, _, _, _] :=
                                                            {rndB, B};
driven[{1, A}, {0, 0}, _, _, _, {4, _}, _, _, _, _, _, _, _] :=
                                                            {rndA, A};
driven[{1, A}, {0, 0}, _, _, _, _, _, _, {2, _}, _, _, _, _] :=
                                                            {rndA, A};
driven[{1, A}, {0, 0}, _, _,_ ,_ ,_ , _, {3, _}, _, _, _] :=
                                                            {rndA, A};
driven[{2, A}, _, {0, 0}, _, _, {3, _}, _, _, _, _, _, _, _] :=
                                                            {rndA, A};
driven[{2, A}, _, {0, 0}, _, _, _, {1, _}, _, _, _, _, _, _] :=
                                                            {rndA, A};
driven[{2, A}, _, {0, 0}, _, _, _, _, _, _, _, {4, _}, _, _] :=
                                                            {rndA, A};
driven[{3, A}, _, _, {0, 0}, _, _, {4, _}, _, _, _, _, _, _] :=
                                                            {rndA, A};
driven[{3, A}, _, _, {0, 0}, _, _, _, {2, _}, _, _, _, _, _] :=
                                                            {rndA, A};
driven[{3, A}, _, _, {0, 0}, _, _, _, _, _, _, _, {1, _}, _] :=
                                                            {rndA, A};
driven[{4, A}, _, _, _, {0, 0}, _, _, {1, _}, _, _, _, _, _] :=
                                                            {rndA, A};
driven[{4, A}, _, _, _, {0, 0}, _, _, _, {3, _}, _, _, _, _] :=
                                                            {rndA, A};
driven[{4, A}, _, _, _, {0, 0}, _, _, _, _, _, _, _, {2, _}] :=
                                                            {rndA, A};
```

```
driven[{1, B}, {0, 0}, _, _, _, {4, _}, _, _, _, _, _, _, _] :=
                                                {rndB, B};
driven[{1, B}, {0, 0}, _, _, _, _, _, _, {2, _}, _, _, _, _] :=
                                                {rndB, B};
driven[{1, B}, {0, 0}, _, _,_ ,_ ,_ ,_ , _, {3, _}, _, _, _] :=
                                                {rndB, B};
driven[{2, B}, _, {0, 0}, _, _, {3, _}, _, _, _, _, _, _, _] :=
                                                {rndB, B};
driven[{2, B}, _, {0, 0}, _, _, _, {1, _}, _, _, _, _, _, _] :=
                                                {rndB, B};
driven[{2, B}, _, {0, 0}, _, _, _, _, _, _, _, {4, _}, _, _] :=
                                                {rndB, B};
driven[{3, B}, _, _, {0, 0}, _, _, {4, _}, _, _, _, _, _, _] :=
                                                {rndB, B};
driven[{3, B}, _, _, {0, 0}, _, _, _, {2, _}, _, _, _, _, _] :=
                                                {rndB, B};
driven[{3, B}, _, _, {0, 0}, _, _, _, _, _, _, _, {1, _}, _] :=
                                                {rndB, B};
driven[{4, B}, _, _, _, {0, 0}, _, _, {1, _}, _, _, _, _, _] :=
                                                {rndB, B};
driven[{4, B}, _, _, _, {0, 0}, _, _, _, {3, _}, _, _, _, _] :=
                                                {rndB, B};
driven[{4, B}, _, _, _, {0, 0}, _, _, _, _, _, _, _, {2, _}] :=
                                                {rndB, B};
driven[{1, _}, {0, 0}, _, _, _, _, _, _, _, _, _, _, _] := {0, 0};
driven[{2, _}, _, {0, 0}, _, _, _, _, _, _, _, _, _, _] := {0, 0};
driven[{3, _}, _, _, {0, 0}, _, _, _, _, _, _, _, _, _] := {0, 0};
driven[{4, _}, _, _, _, {0, 0}, _, _, _, _, _, _, _, _] := {0, 0};
driven[{0, 0}, {3, _}, {4, _}, _, _, _, _, _, _, _, _, _, _]:={0, 0};
driven[{0, 0}, {3, _}, _, {1, _}, _, _, _, _, _, _, _, _, _]:={0, 0};
driven[{0, 0}, {3, _}, _, _, {2, _}, _, _, _, _, _, _, _, _]:={0, 0};
driven[{0, 0}, _, {4, _}, {1, _}, _, _, _, _, _, _, _, _, _]:={0, 0};
driven[{0, 0}, _, {4, _}, _, {2, _}, _, _, _, _, _, _, _, _]:={0, 0};
driven[{0, 0}, _, _, {1, _}, {2, _}, _, _, _, _, _, _, _, _]:={0, 0};
driven[{0, 0}, {3, A}, _, _, _, _, _, _, _, _, _, _, _] := {rndA, A};
driven[{0, 0}, _, {4, A}, _, _, _, _, _, _, _, _, _, _] := {rndA, A};
driven[{0, 0}, _, _, {1, A}, _, _, _, _, _, _, _, _, _] := {rndA, A};
driven[{0, 0}, _, _, _, {2, A}, _, _, _, _, _, _, _, _] := {rndA, A};
driven[{0, 0}, {3, B}, _, _, _, _, _, _, _, _, _, _, _] := {rndB, B};
driven[{0, 0}, _, {4, B}, _, _, _, _, _, _, _, _, _, _] := {rndB, B};
driven[{0, 0}, _, _, {1, B}, _, _, _, _, _, _, _, _, _] := {rndB, B};
driven[{0, 0}, _, _, _, {2, B}, _, _, _, _, _, _, _, _] := {rndB, B};
driven[{0, 0}, _, _, _, _, _, _, _, _, _, _, _, _] := {0, 0};
```

```
    MvonN[func__, lat_] :=
      MapThread[func, Map[RotateRight[lat, #]&,
                  {{0, 0}, {1, 0}, {0, -1}, {-1, 0}, {0, 1},
                   {1, -1}, {-1, -1}, {-1, 1}, {1, 1},
                   {2, 0}, {0, -2}, {-2, 0}, {0, 2}}], 2];


    NestList[MvonN[driven, #]&, initPhase, t]}
    ]
```

Coalescence

```
    stormyWeather[n_, p_, t_]:=
    Module[{RND, cloud, moveCoalesce, VonNeumann},

      RND := Random[Integer, {1, 4}];

      cloud = Table[{ RND , 1} * Floor[Random[] + p] , {n}, {n}];

      moveCoalesce[_, {3, a_}, {4, b_}, {1, c_}, {2, d_}] :=
                                                {RND, a + b + c + d};

      moveCoalesce[_, {3, a_}, _, {1, c_}, {2, d_}] := {RND, a + c + d};
      moveCoalesce[_, _, {4, b_}, {1, c_}, {2, d_}] := {RND, b + c + d};
      moveCoalesce[_, {3, a_}, {4, b_}, {1, c_}, _] := {RND, a + b + c};
      moveCoalesce[_, {3, a_}, {4, b_}, _, {2, d_}] := {RND, a + b + d};

      moveCoalesce[_, _, _, {1, c_}, {2, d_}] := {RND, c + d};
      moveCoalesce[_, {3, a_}, _, {1, c_}, _] := {RND, a + c};
      moveCoalesce[_, _, {4, b_}, {1, c_}, _] := {RND, b + c};
      moveCoalesce[_, {3, a_}, _, _, {2, d_}] := {RND, a + d};
      moveCoalesce[_, _, {4, b_}, _, {2, d_}] := {RND, b + d};
      moveCoalesce[_, {3, a_}, {4, b_}, _, _] := {RND, a + b};

      moveCoalesce[_, _, _, {1, c_}, _] := {RND, c};
      moveCoalesce[_, _, _, _, {2, d_}] := {RND, d};
      moveCoalesce[_, {3, a_}, _, _, _] := {RND, a};
      moveCoalesce[_, _, {4, b_}, _, _] := {RND, b};

      moveCoalesce[_, _, _, _, _] := {0, 0};

      VonNeumann[func__, lat_] :=
        MapThread[func, Map[RotateRight[lat, #]&,
                  {{0, 0}, {1, 0}, {0, -1}, {-1, 0}, {0, 1}}], 2];

      Nest[VonNeumann[moveCoalesce, #]&, cloud, t]
    ]
```

A Model

```
Amodel[p_, n_, t_] :=
 Module[{initConf, ADR, VonNeumann},

  initConf =
        ReplacePart[Table[1, {2n + 1}, {2n + 1}], 0, {n + 1, n + 1}];

  ADR[1, 1, 1, 1, 1] := 1;
  ADR[_, _, _, _, _] := Floor[Random[] + p];

  VonNeumann[func__, lat_] :=
      MapThread[func, Map[RotateRight[lat, #]&,
                 {{0, 0}, {1, 0}, {0, -1}, {-1, 0}, {0, 1}}], 2];

  Nest[VonNeumann[ADR, #]&, initConf, t]
  ]
```

CPM Model

```
ContactProcessModel[n_, p_, q_, t_]:=
 Module[{initConf, CPM, VonNeumann},

  initConf =
        ReplacePart[Table[1, {2n + 1}, {2n + 1}], 0, {n + 1, n + 1}];

  CPM[0, _, _, _, _] := Floor[Random[] + p];
  CPM[1, a_, b_, c_, d_] :=
                    Floor[Random[] + 1 - q * (4 - (a + b + c + d))/4];
  CPM[1, 1, 1, 1, 1] := 1;

  VonNeumann[func__, lat_] :=
      MapThread[func, Map[RotateRight[lat, #]&,
                 {{0, 0}, {1, 0}, {0, -1}, {-1, 0}, {0, 1}}], 2];

  Nest[VonNeumann[CPM, #]&, initConf, t]
  ]
```

Adsorption with Suface Diffusion

```
ADwithDiffusion[p_, n_, r_, t_] :=
 Module[{initConf, ADR, RND, diffuse, VonNeumann, MvonN},

  initConf =
        Table[Random[Integer, {1, 4}] * Floor[p + Random[]], {n}, {n}];
```

```
ADR[_, _, _, _, _] := Random[Integer, {1, 4}] * Floor[Random[] + p];
ADR[_?Positive, _?Positive, _?Positive, _?Positive, _?Positive] :=
                                        Random[Integer, {1, 4}];
RND := Random[Integer, {1, 4}];

diffuse[1, 0, _, _, _, 4, _, _, _, _, _, _, _] := RND;
diffuse[1, 0, _, _, _, _, _, _, 2, _, _, _, _] := RND;
diffuse[1, 0, _, _, _, _, _, _, _, 3, _, _, _] := RND;
diffuse[1, 0, _, _, _, _, _, _, _, _, _, _, _] := 0;
diffuse[2, _, 0, _, _, 3, _, _, _, _, _, _, _] := RND;
diffuse[2, _, 0, _, _, _, 1, _, _, _, _, _, _] := RND;
diffuse[2, _, 0, _, _, _, _, _, _, 4, _, _, _] := RND;
diffuse[2, _, 0, _, _, _, _, _, _, _, _, _, _] := 0;
diffuse[3, _, _, 0, _, _, 4, _, _, _, _, _, _] := RND;
diffuse[3, _, _, 0, _, _, _, 2, _, _, _, _, _] := RND;
diffuse[3, _, _, 0, _, _, _, _, _, _, 1, _, _] := RND;
diffuse[3, _, _, 0, _, _, _, _, _, _, _, _, _] := 0;
diffuse[4, _, _, _, 0, _, _, 1, _, _, _, _, _] := RND;
diffuse[4, _, _, _, 0, _, _, _, 3, _, _, _, _] := RND;
diffuse[4, _, _, _, 0, _, _, _, _, _, _, _, 2] := RND;
diffuse[4, _, _, _, 0, _, _, _, _, _, _, _, _] := 0;
diffuse[_?Positive, _, _, _, _, _, _, _, _, _, _, _, _] := RND;
diffuse[0, 3, 4, _, _, _, _, _, _, _, _, _, _] := 0;
diffuse[0, 3, _, 1, _, _, _, _, _, _, _, _, _] := 0;
diffuse[0, 3, _, _, 2, _, _, _, _, _, _, _, _] := 0;
diffuse[0, _, 4, 1, _, _, _, _, _, _, _, _, _] := 0;
diffuse[0, _, 4, _, 2, _, _, _, _, _, _, _, _] := 0;
diffuse[0, _, _, 1, 2, _, _, _, _, _, _, _, _] := 0;
diffuse[0, 3, _, _, _, _, _, _, _, _, _, _, _] := RND;
diffuse[0, _, 4, _, _, _, _, _, _, _, _, _, _] := RND;
diffuse[0, _, _, 1, _, _, _, _, _, _, _, _, _] := RND;
diffuse[0, _, _, _, 2, _, _, _, _, _, _, _, _] := RND;
diffuse[0, _, _, _, _, _, _, _, _, _, _, _, _] := 0;


VonNeumann[func__, lat_] :=
    MapThread[func, Map[RotateRight[lat, #]&,
            {{0, 0}, {1, 0}, {0, -1}, {-1, 0}, {0, 1}}], 2];


MvonN[func__, lat_] :=
  MapThread[func, Map[RotateRight[lat, #]&,
            {{0, 0}, {1, 0}, {0, -1}, {-1, 0}, {0, 1},
             {1, -1}, {-1, -1}, {-1, 1}, {1, 1},
             {2, 0}, {0, -2}, {-2, 0}, {0, 2}}], 2];


NestList[Nest[MvonN[diffuse, #]&, VonNeumann[ADR, #], r]&,
        initConf, t]
]
```

Chemotaxis

```
chemotaxis[m_, p_, t_] :=
Module[{initConf, sniff, RND, walk, VonNeumann, MvonN},

 initConf =
  Table[{0, Random[Integer, {1, 4}]} * Floor[Random[] + p], {n}, {n}];

 sniff[{a_, 0}, _, _, _, _] := {a, 0};
 sniff[{a_, _}, {n_, _}, {e_, _}, {s_, _}, {w_, _}] :=
  {a, #[[Random[Integer, {1, Length[#]}]]]}&[Flatten[Position[#,
                                             Max[#]]]]&[{n, e, s, w}]];


 walk[{a_, 1}, {_, 0}, _, _, _, {_, 4}, _, _, _, _, _, _, _] := {a, 1};
 walk[{a_, 1}, {_, 0}, _, _, _, _, _, _, {_, 2}, _, _, _, _] := {a, 1};
 walk[{a_, 1}, {_, 0}, _, _, _, _, _, _, _, {_, 3}, _, _, _] := {a, 1};
 walk[{a_, 1}, {_, 0}, _, _, _, _, _, _, _, _, _, _, _] := {a + 1, 0};
 walk[{a_, 2}, _, {_, 0}, _, _, {_, 3}, _, _, _, _, _, _, _] := {a, 2};
 walk[{a_, 2}, _, {_, 0}, _, _, _, {_, 1}, _, _, _, _, _, _] := {a, 2};
 walk[{a_, 2}, _, {_, 0}, _, _, _, _, _, _, _, {_, 4}, _, _] := {a, 2};
 walk[{a_, 2}, _, {_, 0}, _, _, _, _, _, _, _, _, _, _] := {a + 1, 0};
 walk[{a_, 3}, _, _, {_, 0}, _, _, {_, 4}, _, _, _, _, _, _] := {a, 3};
 walk[{a_, 3}, _, _, {_, 0}, _, _, _, {_, 2}, _, _, _, _, _] := {a, 3};
 walk[{a_, 3}, _, _, {_, 0}, _, _, _, _, _, _, {_, 1}, _, _] := {a, 3};
 walk[{a_, 3}, _, _, {_, 0}, _, _, _, _, _, _, _, _, _] := {a + 1, 0};
 walk[{a_, 4}, _, _, _, {_, 0}, _, _, {_, 1}, _, _, _, _, _] := {a, 4};
 walk[{a_, 4}, _, _, _, {_, 0}, _, _, _, {_, 3}, _, _, _, _] := {a, 4};
 walk[{a_, 4}, _, _, _, {_, 0}, _, _, _, _, _, _, _, {_, 2}] := {a, 4};
 walk[{a_, 4}, _, _, _, {_, 0}, _, _, _, _, _, _, _, _] := {a + 1, 0};
 walk[{a_, b_}, _, _, _, _, _, _, _, _, _, _, _, _] := {a + 1, b};
 walk[{a_, 0}, {_, 3}, {_, 4}, _, _, _, _, _, _, _, _, _, _] :=
                                             {Max[a - 1, 0], 0};
 walk[{a_, 0}, {_, 3}, _, {_, 1}, _, _, _, _, _, _, _, _, _] :=
                                             {Max[a - 1, 0], 0};
 walk[{a_, 0}, {_, 3}, _, _, {_, 2}, _, _, _, _, _, _, _, _] :=
                                             {Max[a - 1, 0], 0};
 walk[{a_, 0}, _, {_, 4}, {_, 1}, _, _, _, _, _, _, _, _, _] :=
                                             {Max[a - 1, 0], 0};
 walk[{a_, 0}, _, {_, 4}, _, {_, 2}, _, _, _, _, _, _, _, _] :=
                                             {Max[a - 1, 0], 0};
 walk[{a_, 0}, _, _, {_, 1}, {_, 2}, _, _, _, _, _, _, _, _] :=
                                             {Max[a - 1, 0], 0};
 walk[{a_, 0}, {_, 3}, _, _, _, _, _, _, _, _, _, _, _] := {a, 3};
 walk[{a_, 0}, _, {_, 4}, _, _, _, _, _, _, _, _, _, _] := {a, 4};
 walk[{a_, 0}, _, _, {_, 1}, _, _, _, _, _, _, _, _, _] := {a, 1};
```

```
walk[{a_, 0}, _, _, _, {_, 2}, _, _, _, _, _, _, _, _] := {a, 2};
walk[{a_, 0}, _, _, _, _, _, _, _, _, _, _, _, _] :=
                                            {Max[a - 1, 0], 0};


   VonNeumann[func__, lat_] :=
       MapThread[func, Map[RotateRight[lat, #]&,
              {{0, 0}, {1, 0}, {0, -1}, {-1, 0}, {0, 1}}], 2];


   MvonN[func__, lat_] :=
     MapThread[func, Map[RotateRight[lat, #]&,
              {{0, 0}, {1, 0}, {0, -1}, {-1, 0}, {0, 1},
               {1, -1}, {-1, -1}, {-1, 1}, {1, 1},
               {2, 0}, {0, -2}, {-2, 0}, {0, 2}}], 2];


   NestList[MvonN[walk, #]&[VonNeumann[sniff, #]]&, initConf, t]
   ]
```

Ant Colony Activity

```
maggiesFarm[n_, p_, s_, r_, t_]:=
Module[{antPopulation, antFarm, border, ant, MvonN},

  antPopulation =
  Table[{{-1, 1, 2, 3, 4}[[Random[Integer, {1, 5}] ]],
       Random[Integer, {1, s}]}]* Floor[Random[] + p],
              {n - 1}, {n - 1}] /. {-1, _} -> {-1, 0};

  border =
   Append[Map[Append[#, {b, 0}]&, #], Table[{b, 0}, {Length[#] + 1}]]&;

  antFarm = border[antPopulation];

  RND := Random[Integer, {1, 4}];

  ant[{-1,   0}, {x_?Positive, _}, _, _, _, _, _, _, _, _, _, _, _] :=
                                                         {RND, 1};
  ant[{-1,   0},   _, {x_?Positive,_}, _, _, _, _, _, _, _, _, _] :=
                                                         {RND, 1};
  ant[{-1,   0},   _, _, {x_?Positive,_}, _, _, _, _, _, _, _, _] :=
                                                         {RND, 1};
  ant[{-1,   0},   _, _, _, {x_?Positive,_ , _, _, _, _, _, _, _] :=
                                                         {RND, 1};
  ant[{-1, 0}, _, _, _, _, _, _, _, _, _, _, _, _] :=
          {-1 + # * Random[Integer, {2, 5}], #}&[Floor[Random[] + r]];
  ant[{x_?Positive, s}, _, _, _, _, _, _, _, _, _, _, _, _] := {-1, 0};
```

```
ant[{1, y_}, {-1 | x_?Positive | b, _}, _, _, _, _,
                                _, _, _, _, _, _, _] := {RND, y + 1};
ant[{2, y_}, _, {-1 | x_?Positive | b, _}, _, _, _,
                                _, _, _, _, _, _, _] := {RND, y + 1};
ant[{3, y_}, _, _, {-1 | x_?Positive | b, _}, _, _,
                                _, _, _, _, _, _, _] := {RND, y + 1};
ant[{4, y_}, _, _, _, {-1 | x_?Positive | b, _}, _,
                                _, _, _, _, _, _, _] := {RND, y + 1};
ant[{1, y_}, {0, 0}, _, _, _, {4, _}, _, _, _, _, _, _, _] :=
                                {RND, y + 1};
ant[{1, y_}, {0, 0}, _, _, _, _, _, _, {2, _}, _, _, _, _] :=
                                {RND, y + 1};
ant[{1, y_}, {0, 0}, _, _,_ ,_ ,_ ,_ , _, {3, _}, _, _, _] :=
                                {RND, y + 1};
ant[{2, y_}, _, {0, 0}, _, _, {3, _}, _, _, _, _, _, _, _] :=
                                {RND, y + 1};
ant[{2, y_}, _, {0, 0}, _, _, _, {1, _}, _, _, _, _, _, _] :=
                                {RND, y + 1};
ant[{2, y_}, _, {0, 0}, _, _, _, _, _, _, _, {4, _}, _, _] :=
                                {RND, y + 1};
ant[{3, y_}, _, _, {0, 0}, _, _, {4, _}, _, _, _, _, _, _] :=
                                {RND, y + 1};
ant[{3, y_}, _, _, {0, 0}, _, _, _, {2, _}, _, _, _, _, _] :=
                                {RND, y + 1};
ant[{3, y_}, _, _, {0, 0}, _, _, _, _, _, _, _, {1, _}, _] :=
                                {RND, y + 1};
ant[{4, y_}, _, _, _, {0, 0}, _, _, {1, _}, _, _, _, _, _] :=
                                {RND, y + 1};
ant[{4, y_}, _, _, _, {0, 0}, _, _, _, {3, _}, _, _, _, _] :=
                                {RND, y + 1};
ant[{4, y_}, _, _, _, {0, 0}, _, _, _, _, _, _, _, {2, _}] :=
                                {RND, y + 1};
ant[{1, _}, {0, 0}, _, _, _, _, _, _, _, _, _, _, _] := {0, 0};
ant[{2, _}, _, {0, 0}, _, _, _, _, _, _, _, _, _, _] := {0, 0};
ant[{3, _}, _, _, {0, 0}, _, _, _, _, _, _, _, _, _] := {0, 0};
ant[{4, _}, _, _, _, {0, 0}, _, _, _, _, _, _, _, _] := {0, 0};
ant[{0, 0}, {3, _}, {4, _}, _, _, _, _, _, _, _, _, _, _] := {0, 0};
ant[{0, 0}, {3, _}, _, {1, _}, _, _, _, _, _, _, _, _, _] := {0, 0};
ant[{0, 0}, {3, _}, _, _, {2, _}, _, _, _, _, _, _, _, _] := {0, 0};
ant[{0, 0}, _, {4, _}, {1, _}, _, _, _, _, _, _, _, _, _] := {0, 0};
ant[{0, 0}, _, {4, _}, _, {2, _}, _, _, _, _, _, _, _, _] := {0, 0};
ant[{0, 0}, _, _, {1, _}, {2, _}, _, _, _, _, _, _, _, _] := {0, 0};
ant[{0, 0}, {3, y_?(# < s &)}, _, _, _, _, _, _, _, _, _, _, _] :=
                                {RND, y + 1};
ant[{0, 0}, _, {4, y_?(# < s &)}, _, _, _, _, _, _, _, _, _, _] :=
                                {RND, y + 1};
```

```
ant[{0, 0}, _, _, {1, y_?(# < s &)}, _, _, _, _, _, _, _, _, _] :=
                                                        {RND, y + 1};
ant[{0, 0}, _, _, _, {2, y_?(# < s &)}, _, _, _, _, _, _, _, _] :=
                                                        {RND, y + 1};
ant[{0, 0}, _, _, _, _, _, _, _, _, _, _, _, _] := {0, 0};
ant[{b, 0}, _, _, _, _, _, _, _, _, _, _, _, _] := {b, 0};


MvonN[func__, lat_] :=
 MapThread[func, Map[RotateRight[lat, #]&,
          {{0, 0}, {1, 0}, {0, -1}, {-1, 0}, {0, 1},
           {1, -1}, {-1, -1}, {-1, 1}, {1, 1},
           {2, 0}, {0, -2}, {-2, 0}, {0, 2}}], 2];


NestList[MvonN[ant, #]&, antFarm, t];
]
```

Grazing Herd Ecosystem

```
predatorPrey[n_, preyDensity_, predDensity_, preg_,
            starve_, p_, t_] /; preyDensity + predDensity < 1 :=
Module[{eco, RND, pasture MvonN},

pasture =
 Table[Floor[Random[] + (preyDensity + predDensity)] *
       Floor[1 + Random[] + predDensity/(preyDensity + predDensity)],
       {n}, {n}] /.
2 :> {RND, Random[Integer, {0, preg}], Random[Integer, {1, starve}]};

RND:=Random[Integer,{1,4}];

eco[{_, 0, 0}, _, _, _, _, _, _, _, _, _, _, _] :=
                                            {RND, preg, starve};
eco[{_, _, 0}, _, _, _, _, _, _, _, _, _, _, _] := 0;
eco[{1, a_, b_?Positive}, 0 | 1, _, _, _, {4, _, _?Positive},
                  _, _, _, _, _, _, _] := {RND, Max[0, a - 1], b - 1};
eco[{1, a_, b_?Positive}, 0 | 1, _, _, _, _, _, _,
     {2, _, _?Positive}, _, _, _, _] := {RND, Max[0, a - 1], b - 1};
eco[{1, a_, b_?Positive}, 0 | 1, _, _, _, _, _, _,
         {3, _, _?Positive}, _, _, _] := {RND, Max[0, a - 1], b - 1};
eco[{2, a_, b_?Positive}, _, 0 | 1, _, _, {3, _, _?Positive},
                  _, _, _, _, _, _, _] := {RND, Max[0, a - 1], b - 1};
eco[{2, a_, b_?Positive}, _, 0 | 1, _, _, _, {1, _, _?Positive},
                  _, _, _, _, _, _] := {RND, Max[0, a - 1], b - 1};
eco[{2, a_, b_?Positive}, _, 0 | 1, _, _, _, _, _,
     _, _, {4, _, _?Positive}, _, _] := {RND, Max[0, a - 1], b - 1};
```

```
eco[{3, a_, b_?Positive}, _, _, 0 | 1, _, _, {4, _, _?Positive},
                _, _, _, _, _, _] := {RND, Max[0, a - 1], b - 1};
eco[{3, a_, b_?Positive}, _, _, 0 | 1, _, _, _,
   {2, _, _?Positive}, _, _, _, _, _] := {RND, Max[0, a - 1], b - 1};
eco[{3, a_, b_?Positive}, _, _, 0 | 1, _, _, _, _,
      _, _, _, {1, _, _?Positive}, _] := {RND, Max[0, a - 1], b - 1};
eco[{4, a_, b_?Positive}, _, _, _, 0 | 1, _, _,
   {1, _, _?Positive}, _, _, _, _, _] := {RND, Max[0, a - 1], b - 1};
eco[{4, a_, b_?Positive}, _, _, _, 0 | 1, _, _, _,
       {3, _, _?Positive}, _, _, _, _] := {RND, Max[0, a - 1], b - 1};
eco[{4, a_, b_?Positive}, _, _, _, 0 | 1, _, _, _,
      _, _, _, _, {2, _, _?Positive}] := {RND, Max[0, a - 1], b - 1};
eco[1, {3, _, _?Positive}, {4, _, _?Positive},
                         _, _, _, _, _, _, _, _, _] := 1;
eco[1, {3, _, _?Positive}, _, {1, _, _?Positive},
                         _, _, _, _, _, _, _, _] := 1;
eco[1, {3, _, _?Positive}, _, _, {2, _, _?Positive},
                         _, _, _, _, _, _, _] := 1;
eco[1, _, {4, _, _?Positive}, {1, _, _?Positive},
                         _, _, _, _, _, _, _, _] := 1;
eco[1, _, {4, _, _?Positive}, _, {2, _, _?Positive},
                         _, _, _, _, _, _, _] := 1;
eco[1, _, _, {1, _, _?Positive}, {2, _, _?Positive},
                         _, _, _, _, _, _, _] := 1;
eco[0, {3, _, _?Positive}, {4, _, _?Positive},
          _, _, _, _, _, _, _, _, _] := Floor[p + Random[]];
eco[0, {3, _, _?Positive}, _, {1, _, _?Positive},
          _, _, _, _, _, _, _, _] := Floor[p + Random[]];
eco[0, {3, _, _?Positive}, _, _, {2, _, _?Positive},
             _, _, _, _, _, _, _] := Floor[p + Random[]];
eco[0, _, {4, _, _?Positive}, {1, _, _?Positive},
          _, _, _, _, _, _, _, _] := Floor[p + Random[]];
eco[0, _, {4, _, _?Positive}, _, {2, _, _?Positive},
             _, _, _, _, _, _, _] := Floor[p + Random[]];
eco[0, _, _, {1, _, _?Positive}, {2, _, _?Positive},
             _, _, _, _, _, _, _] := Floor[p + Random[]];
eco[{1, 0, _?Positive}, 0 | 1, _, _, _, _, _, _, _, _, _, _] :=
                                         {RND, preg, starve};
eco[{2, 0, _?Positive}, _, 0 | 1, _, _, _, _, _, _, _, _, _] :=
                                         {RND, preg, starve};
eco[{3, 0, _?Positive}, _, _, 0 | 1, _, _, _, _, _, _, _, _] :=
                                         {RND, preg, starve};
eco[{4, 0, _?Positive}, _, _, _, 0 | 1, _, _, _, _, _, _, _] :=
                                         {RND, preg, starve};
eco[{1, _, _?Positive}, 0 | 1, _, _, _, _, _, _, _, _, _, _] := 0;
eco[{2, _, _?Positive}, _, 0 | 1, _, _, _, _, _, _, _, _, _] := 0;
eco[{3, _, _?Positive}, _, _, 0 | 1, _, _, _, _, _, _, _, _] := 0;
```

```
eco[{4, _, _?Positive}, _, _, _, 0 | 1, _, _, _, _, _, _, _, _] := 0;
eco[1, {3, 0, _?Positive}, _, _, _, _, _, _, _, _, _, _, _] :=
                                              {RND, preg, starve};
eco[1, _, {4, 0, _?Positive}, _, _, _, _, _, _, _, _, _, _] :=
                                              {RND, preg, starve};
eco[1, _, _, {1, 0, _?Positive}, _, _, _, _, _, _, _, _, _] :=
                                              {RND, preg, starve};
eco[1, _, _, _, {2, 0, _?Positive}, _, _, _, _, _, _, _, _] :=
                                              {RND, preg, starve};
eco[0, {3, 0, b_?Positive}, _, _, _, _, _, _, _, _, _, _, _] :=
                                              {RND, preg, b - 1};
eco[0, _, {4, 0, b_?Positive}, _, _, _, _, _, _, _, _, _, _] :=
                                              {RND, preg, b - 1};
eco[0, _, _, {1, 0, b_?Positive}, _, _, _, _, _, _, _, _, _] :=
                                              {RND, preg, b - 1};
eco[0, _, _, _, {2, 0, b_?Positive}, _, _, _, _, _, _, _, _] :=
                                              {RND, preg, b - 1};
eco[1, {3, a_, _?Positive}, _, _, _, _, _, _, _, _, _, _, _] :=
                                              {RND, a -1, starve};
eco[1, _, {4, a_, _?Positive}, _, _, _, _, _, _, _, _, _, _] :=
                                              {RND, a -1, starve};
eco[1, _, _, {1, a_, _?Positive}, _, _, _, _, _, _, _, _, _] :=
                                              {RND, a -1, starve};
eco[1, _, _, _, {2, a_, _?Positive}, _, _, _, _, _, _, _, _] :=
                                              {RND, a -1, starve};
eco[0, {3, a_, b_?Positive}, _, _, _, _, _, _, _, _, _, _, _] :=
                                              {RND, a - 1, b - 1};
eco[0, _, {4, a_, b_?Positive}, _, _, _, _, _, _, _, _, _, _] :=
                                              {RND, a - 1, b - 1};
eco[0, _, _, {1, a_, b_?Positive}, _, _, _, _, _, _, _, _, _] :=
                                              {RND, a - 1, b - 1};
eco[0, _, _, _, {2, a_, b_?Positive}, _, _, _, _, _, _, _, _] :=
                                              {RND, a - 1, b - 1};
eco[{_, a_, b_}, _, _, _, _, _, _, _, _, _, _, _, _] :=
                                              {RND, a - 1, b - 1};
eco[1, _, _, _, _, _, _, _, _, _, _, _, _] := 1;
eco[0, _, _, _, _, _, _, _, _, _, _, _, _] := Floor[p + Random[]];


MvonN[func__, lat_] :=
 MapThread[func, Map[RotateRight[lat, #]&,
          {{0, 0}, {1, 0}, {0, -1}, {-1, 0}, {0, 1},
           {1, -1}, {-1, -1}, {-1, 1}, {1, 1},
           {2, 0}, {0, -2}, {-2, 0}, {0, 2}}], 2];


NestList[MvonN[eco, #]&, pasture, t]
]
```

Deterministic Epidemic

```
contagion[n_, s_, a_, g_, t_] :=
 Module[{population, spread, VonNeumann},

   population = Table[Floor[1 + s - Random[]] *
                     Random[Integer, {1, a + g}] , {n}, {n}];

   spread[ 0 | (a + g), _, _, _, _ ] := 0;
   spread[x_?Positive, _, _, _, _] := x + 1;
   spread[0, u_, v_, w_, x_] := 1 /; MemberQ[Range[a], u | v | w | x];

   VonNeumann[func__, lat_] :=
      MapThread[func, Map[RotateRight[lat, #]&,
               {{0, 0}, {1, 0}, {0, -1}, {-1, 0}, {0, 1}}], 2];

   NestList[VonNeumann[spread, #]&, population, t]
 ]
```

Stochastic Forest Fire

```
forestFire[n_, s_, k_, p_, f_, g_, t_] :=
 Module[{ForestPreserve, spread, sprout, catch, spont, VonNeumann},

   sprout = (1 + p);
   catch = (2 - g);
   spont = 1 + f (1- g);

   forestPreserve = Table[Floor[1 + s - Random[]], {n}, {n}] /.
                            1 :> Floor[1 + k + Random[]];

   spread[0, _, _, _, _] := Floor[sprout - Random[] ];
   spread[2, _, _, _, _] = 0;
   spread[1, a_, b_, c_, d_] :=
          1 + Floor[catch - Random[] ] /; MatchQ[2, a | b | c | d];
   spread[1, a_, b_, c_, d_] := 1 + Floor[spont - Random[]];

   VonNeumann[func__, lat_] :=
      MapThread[func, Map[RotateRight[lat, #]&,
               {{0, 0}, {1, 0}, {0, -1}, {-1, 0}, {0, 1}}], 2];

   NestList[VonNeumann[spread, #]&, forestPreserve, t]
 ]
```

Hodgepodge

```
hodgepodge[r_Integer, s_Integer, k1_, k2_, g_Integer, t_Integer] :=
Module[{initconfig, VonNeumann, sick},

  initconfig = Table[Random[Integer, {0, r}], {s}, {s}];

  sick[r, _, _, _, _]  := 0;

  sick[0, b_, c_, d_, e_] :=
    Min[r, Floor[(Floor[N[b/r]] + Floor[N[c/r]]  +
                   Floor[N[d/r]]  + Floor[N[e/r]])/k1] +
            Floor[(Sign[Mod[b, r]] + Sign[Mod[c, r]] +
                   Sign[Mod[d, r]] + Sign[Mod[e, r]])/k2]
        ];

  sick[a_, b_, c_, d_, e_] :=
    Min[r, g + Floor[(a + b + c + d + e)/
                    (Sign[Mod[a, r]] + Sign[Mod[b, r]] + Sign[Mod[c, r]]
                    + Sign[Mod[d, r]] + Sign[Mod[e, r]])]
    ];

  VonNeumann[func__, lat_] :=
    MapThread[func, Map[RotateRight[lat, #]&,
            {{0, 0}, {1, 0}, {0, -1}, {-1, 0}, {0, 1}}], 2];

  FixedPoint[VonNeumann[sick, #]&, initconfig, t]
]
```

Sandpile

```
catastrophe[s_, m_]:=
Module[{absorbBC, landscape, topple, VonNeumann},

  absorbBC = (Prepend[Append[Map[Prepend[Append[#, 0], 0]&, #],
                              Table[0, {Length[#] + 2}]],
                      Table[0, {Length[#] + 2}]])&;

  landscape = absorbBC[Table[Random[Integer, {1, 4}], {s}, {s}]];

  While[Max[landscape]  <  5,
        randx = Random[Integer, {2, s + 1}, {2, s + 1}];
        randy = Random[Integer, {2, s + 1}, {2, s + 1}];
        landscape[[randx, randy]]++
        ];
```

```
topple[0, _, _, _, _ ] := 0;
topple[a_ /; a < 5 , b_, c_, d_, e_] :=
            a + Floor[b/5] + Floor[c/5] + Floor[d/5] + Floor[e/5];
topple[a_, b_, c_, d_, e_] :=
            a - 4 + Floor[b/5] + Floor[c/5] + Floor[d/5] + Floor[e/5];


VonNeumann[func__, lat_] :=
  MapThread[func, Map[RotateRight[lat, #]&,
            {{0, 0}, {1, 0}, {0, -1}, {-1, 0}, {0, 1}}], 2];


FixedPointList[VonNeumann[topple, #]&, landscape, m]
]
```

Spatial Prisoner's Dilemma

```
spatialPrisonersDilemma[n_, p_,  t_] :=
Module[{initConf, Moore, totalPayoff},

initConf =
  ReplacePart[Table[1, {2 n + 1}, {2 n + 1}], 0, {n + 1, n + 1}];

totalPayoff[1, a_, b_, c_, d_, e_, f_, g_, h_] :=
                        {1 + a + b + c + d + e + f + g + h, 1};
totalPayoff[0, a_, b_, c_, d_, e_, f_, g_, h_] :=
                        {p (a + b + c + d + e + f + g + h), 0};

Moore[func__, lat_] :=
  MapThread[func, Map[RotateRight[lat, #]&,
            {{0, 0}, {1, 0}, {0, -1}, {-1, 0}, {0, 1},
            {1, -1}, {-1, -1}, {-1, 1}, {1, 1}}], 2];

NestList[Moore[Last[Sort[{##}]][[2]]&, #]&[Moore[totalPayoff, #]]&,
         initConf, t]
]
```

Index

MODELING NATURE
Cellular Automata Simulations with *Mathematica*®

Since this field is fast-moving, we expect updates and changes to occur that might necessitate sending you the most current pertinent information by paper, electronic media, or both, regarding *Modeling Nature: Cellular Automata Simulations with Mathematica*®. Therefore, in order to not miss out on receiving your important update information, please fill out this card and return it to us promptly. Thank you.

Name: _____

Title: _____

Company: _____

Address: _____

City: _____ State: _____ Zip: _____

Country: _____ Phone: _____

E-mail: _____

Areas of Interest/Technical Expertise: _____

Comments on this Publication: _____

_____

_____

☐ Please check this box to indicate that we may use your comments in our promotion and advertising for this publication.

Purchased from: _____
Date of Purchase: _____

☐ Please add me to your mailing list to receive updated information on *Modeling Nature: Cellular Automata Simulations with Mathematica*® and other TELOS publications.

☐ I have a   ☐ IBM compatible   ☐ Macintosh   ☐ UNIX   ☐ other

Designate specific model _____

PLEASE TAPE HERE

FOLD HERE

- -

NO POSTAGE
NECESSARY
IF MAILED
IN THE
UNITED STATES

BUSINESS REPLY MAIL
FIRST CLASS MAIL PERMIT NO. 1314 SANTA CLARA, CA

POSTAGE WILL BE PAID BY ADDRESSEE

TELOS
THE
ELECTRONIC
LIBRARY
OF
SCIENCE

3600 PRUNERIDGE AVE STE 200
SANTA CLARA CA 95051-9835